Electrophoresis

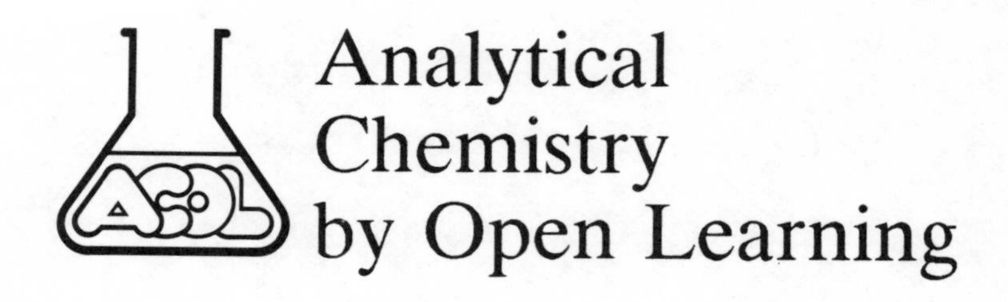
Analytical
Chemistry
by Open Learning

Titles in Series:

Samples and Standards
Sample Pretreatment
Classical Methods
Measurement, Statistics and Computation
Using Literature
Instrumentation
Chromatographic Separations
Gas Chromatography
High Performance Liquid Chromatography
Electrophoresis
Thin Layer Chromatography
Visible and Ultraviolet Spectroscopy
Fluorescence and Phosphorescence Spectroscopy
Infra Red Spectroscopy
Atomic Absorption and Emission Spectroscopy
Nuclear Magnetic Resonance Spectroscopy
X-Ray Methods
Mass Spectrometry
Scanning Electron Microscopy and X-Ray Microanalysis
Principles of Electroanalytical Methods
Potentiometry and Ion Selective Electrodes
Polarography and Other Voltammetric Methods
Radiochemical Methods
Clinical Specimens
Diagnostic Enzymology
Quantitative Bioassay
Assessment and Control of Biochemical Methods
Thermal Methods
Microprocessor Applications

Electrophoresis

Analytical Chemistry by Open Learning

Author:
MAUREEN MELVIN
Robert Gordon's Institute of Technology, Aberdeen

Editor:
DAVID KEALEY

on behalf of ACOL

Published on behalf of ACOL, London
by
JOHN WILEY & SONS
Chichester · New York · Brisbane · Toronto · Singapore

Published by permission of the Controller of
Her Majesty's Stationery Office

Library of Congress Cataloging in Publication Data:

Melvin, Maureen
Electrophoresis

(Analytical chemistry by open learning)
Bibliography: p.
Includes index.
1. Electrophoresis—Programmed instruction.
2. Chemistry, Analytic—Programmed instruction.
I. Kealey, D. (David) II. Title. III. Series.
QD79.E44M45 1987 543'.08 86-28195
ISBN 0 471 91374 X
ISBN 0 471 91375 8 (pbk.)

British Library Cataloguing in Publication Data:

Melvin, Maureen
Electrophoresis.—(Analytical chemistry)
1. Electrophoresis
I. Title II. Kealey, D. III. Analytical Chemistry by Open Learning *(Project)*
IV. Series
541.3'7 QD79.E44

ISBN 0 471 91374 X
ISBN 0 471 91375 8 Pbk

Printed and bound in Great Britain

Analytical Chemistry

This series of texts is a result of an initiative by the Committee of Heads of Polytechnic Chemistry Departments in the United Kingdom. A project team based at Thames Polytechnic using funds available from the Manpower Services Commission 'Open Tech' Project have organised and managed the development of the material suitable for use by 'Distance Learners'. The contents of the various units have been identified, planned and written almost exclusively by groups of polytechnic staff, who are both expert in the subject area and are currently teaching in analytical chemistry.

The texts are for those interested in the basics of analytical chemistry and instrumental techniques who wish to study in a more flexible way than traditional institute attendance or to augment such attendance. A series of these units may be used by those undertaking courses leading to BTEC (levels IV and V), Royal Society of Chemistry (Certificates of Applied Chemistry) or other qualifications. The level is thus that of Senior Technician.

It is emphasised however that whilst the theoretical aspects of analytical chemistry can be studied in this way there is no substitute for the laboratory to learn the associated practical skills. In the U.K. there are nominated Polytechnics, Colleges and other Institutions who offer tutorial and practical support to achieve the practical objectives identified within each text. It is expected that many institutions worldwide will also provide such support.

The project will continue at Thames Polytechnic to support these 'Open Learning Texts', to continually refresh and update the material and to extend its coverage.

Further information about nominated support centres, the material or open learning techniques may be obtained from the project office at Thames Polytechnic, ACOL, Wellington St., Woolwich, London, SE18 6PF.

How to Use an Open Learning Text

Open learning texts are designed as a convenient and flexible way of studying for people who, for a variety of reasons cannot use conventional education courses. You will learn from this text the principles of one subject in Analytical Chemistry, but only by putting this knowledge into practice, under professional supervision, will you gain a full understanding of the analytical techniques described.

To achieve the full benefit from an open learning text you need to plan your place and time of study.

- Find the most suitable place to study where you can work without disturbance.

- If you have a tutor supervising your study discuss with him, or her, the date by which you should have completed this text.

- Some people study perfectly well in irregular bursts, however most students find that setting aside a certain number of hours each day is the most satisfactory method. It is for you to decide which pattern of study suits you best.

- If you decide to study for several hours at once, take short breaks of five or ten minutes every half hour or so. You will find that this method maintains a higher overall level of concentration.

Before you begin a detailed reading of the text, familiarise yourself with the general layout of the material. Have a look at the course contents list at the front of the book and flip through the pages to get a general impression of the way the subject is dealt with. You will find that there is space on the pages to make comments alongside the text as you study—your own notes for highlighting points that you feel are particularly important. Indicate in the margin the points you would like to discuss further with a tutor or fellow student. When you come to revise, these personal study notes will be very useful.

Π When you find a paragraph in the text marked with a symbol such as is shown here, this is where you get involved. At this point you are directed to do things: draw graphs, answer questions, perform calculations, etc. Do make an attempt at these activities. If necessary cover the succeeding response with a piece of paper until you are ready to read on. This is an opportunity for you to learn by participating in the subject and although the text continues by discussing your response, there is no better way to learn than by working things out for yourself.

We have introduced self assessment questions (SAQ) at appropriate places in the text. These SAQs provide for you a way of finding out if you understand what you have just been studying. There is space on the page for your answer and for any comments you want to add after reading the author's response. You will find the author's response to each SAQ at the end of the text. Compare what you have written with the response provided and read the discussion and advice.

At intervals in the text you will find a Summary and List of Objectives. The Summary will emphasise the important points covered by the material you have just read and the Objectives will give you a checklist of tasks you should then be able to achieve.

You can revise the Unit, perhaps for a formal examination, by re-reading the Summary and the Objectives, and by working through some of the SAQs. This should quickly alert you to areas of the text that need further study.

At the end of the book you will find for reference lists of commonly used scientific symbols and values, units of measurement and also a periodic table.

Contents

Study Guide

The technique of electrophoresis is used widely by chemists and biochemists. It has particular importance in the study of environmental health, medical technology, food analysis, pollution control, water analysis and forensic investigations.

This Unit assumes that you have an understanding of chemistry equivalent to that of a student who has passed HNC or HTC in chemistry or BTEC, and a knowledge of physics to at least GCE (OL). You are also expected to have some knowledge of elementary biological chemistry and in particular, to be familiar with the structure and function of amino acids, peptides and proteins, nucleotides, polynucleotides and nucleic acids. To check your knowledge of these topics you should try answering SAQs 1a – 1d which are given at the end of this Study Guide. If you feel that you need to 'brush up' on these aspects of biological chemistry, I recommend that you obtain a copy of one (or more) of the following textbooks from your local library, or from a bookseller:

Conn, E. E. and Stumpf, P. K. *Outlines of Biochemistry*, 4th edn, John Wiley & Sons, 1976.

Hart, H. and Schuetz, R. D. *Organic Chemistry*, 5th edn, Houghton Mifflin Company, 1978.

Lehninger, A. *Biochemistry: the Molecular Basis of Cell Structure and Function*, 2nd edn, Worth, 1976.

Stryer, L. *Biochemistry*, 2nd edn, W. H. Freeman, 1981.

Electrophoresis is a technique based, in part, on chromatographic principles. Electrophoresis, like chromatography, is concerned with the separation of the components of samples. In fact, electrophoresis and chromatography are sometimes used in combination with each other to achieve high resolution in the separation of molecules, for example amino acids and peptides derived from the hydrolysis of proteins. The emphasis in this Unit is on practical aspects of electrophoresis and on its applications in chemical analysis.

The Electrophoresis Unit is divided into seven main sections. Your understanding of the material is checked as you progress through the Unit by means of questions in the text and self assessment questions (SAQs), which have been designed to emphasise the concepts in the Unit that I regard as particularly important. The correct responses to the SAQs are given at the end of the Unit. If you have difficulty in answering any SAQs correctly, you are advised to refer back to the relevant parts of the text.

The aims of this Unit are that you will develop an understanding of the principles of electrophoretic techniques, and a competent working knowledge of the applications of electrophoresis and its importance in analytical chemistry.

Now here are some SAQs that have been designed to test your knowledge of those areas of biological chemistry that I consider a necessary prerequisite for this Unit. If you have difficulty with them, I suggest that you consult the textbooks I mentioned earlier.

When you have attempted each one, turn to the responses which are printed at the end of the Unit.

SAQ 1a

The general structure of the 20 amino acids (more correctly called 1-amino-1-carboxylic acids) commonly found in plant and animal proteins is

$$\begin{array}{l} R-CH-COOH \\ \quad\;\;\; | \\ \quad\; NH_2 \end{array}$$

where R represents one of 20 different groups. (Table 5 at the end of this Unit shows the full structure of these amino acids). Using this general formula, show how three amino acids are linked together to form a tripeptide.

SAQ 1b

Although proline is listed in Table 5 as one of the 20 amino acids commonly found in plant and animal proteins, it is not a 1-amino-1-carboxylic acid. Write down the structural formula of the tripeptide in which proline is the central amino acid, linked to two others of structural formulae

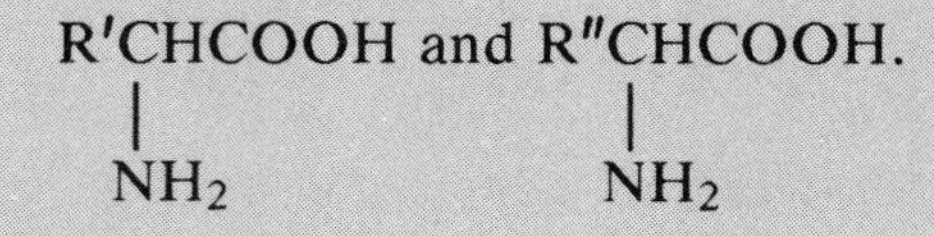

SAQ 1c

Name and draw the structural formulae of the major components produced on complete hydrolysis of (*i*) RNA and (*ii*) DNA.

SAQ 1d

Draw a diagram to illustrate the polynucleotide structure of a single strand of RNA. Indicate how the structure of DNA differs from what you have drawn.

When you feel confident about the material tested in these SAQs, we can proceed with this Unit.

Supporting Practical Work

An appropriate practical course in Electrophoresis is an essential component of any comprehensive study of the subject.

The aims of such a practical course would be to:

- illustrate important principles from the theory part of the Unit.
- provide experience in handling various types of equipment used for electrophoresis.
- allow the opportunity to use a variety of different electrophoretic techniques.
- enable students to select and apply electrophoretic techniques appropriate to particular analytical problems.
- demonstrate the production of different types of electrophoretograms.
- provide a variety of types of electrophoretogram, and illustrate their interpretation.

Parts 1 and 2 of the Electrophoresis Unit comprise mainly introductory and descriptive material, therefore they require no complementary practical component.

Sections 3 and 5 of the Unit offer ample opportunity for associated Practical work. It is important that several different types of electrophoretic support medium and equipment are provided for 'hands-on' experience; in addition, a number of different detection techniques should be carried out or demonstrated.

Section 4 is theoretical, and would not require associated practical experimentation. However, it is suggested that some numerical examples illustrating the principles would be a useful adjunct to this Section.

Students would benefit from experience of, or demonstration of, the techniques in Sections 6 and 7. In the absence of appropriate facilities, 'Data Packs' of experimental results could be prepared, so that opportunity is given for the interpretation and analysis of 'real' electrophoretograms illustrating the techniques described in these Sections.

Bibliography

If you would like to read more about electrophoresis, I recommend the following books:

Gaal, O., Medgyesi, G.A. and Kereczkey, L., *Electrophoresis in the separation of Biological Macromolecules*, John Wiley and Sons (1980)

Andrews, A.T. *Electrophoresis*, Oxford University Press (1986).

1. General Introduction to Electrophoresis

1.1. INTRODUCTION

Electrophoresis as an analytical tool was introduced by the Swedish chemist Arne Tiselius, first in his doctoral thesis in 1930, then later in a modified and improved form (Tiselius, 1937).

Tiselius's main interest was in the chemistry of the serum proteins. His investigations resulted in the development of specialised apparatus and methodology for electrophoresis from which have been derived the techniques we use today. For his pioneer work in this field, Tiselius was awarded the Nobel prize in 1948.

The detailed theory of electrophoresis is very complicated and still incomplete! However, a simple description of the principles is sufficient for understanding most of the common applications of the technique.

Electrophoresis involves the separation of charged species (molecules) on the basis of their movement under the influence of an applied electric field.

Π The fundamental requirement for electrophoresis is that the molecules under study must have either a net positive or negative charge. Why do you think this is necessary?

As shown in Fig. 1.1a, under the influence of an applied electric field, positively charged molecules will move towards the cathode whilst negatively charged molecules move towards the anode. A molecule with no net charge should remain stationary.

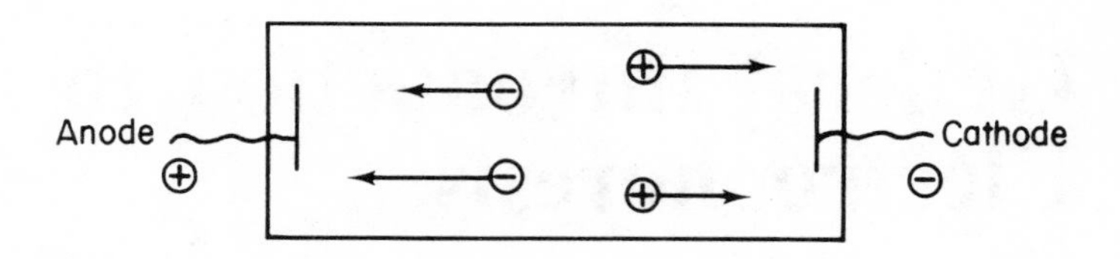

Fig. 1.1a. *Movement of charged molecules under the influence of an applied electric field*

In electrophoresis, the species under study move in a liquid medium which is usually supported by an inert solid substance, such as paper or a semi-solid gel. The liquid serves as a conducting medium for the electric current generated by the application of an external voltage to the system.

The velocity with which a molecule moves towards the anode or cathode during electrophoresis is called the *migration velocity* of the molecule. It is given the symbol v, its units are cm s^{-1}, and its value is a characteristic both of the molecule itself and the strength of the applied electric field. For any molecule migrating in a particular liquid, medium its migration velocity (v) will increase with increasing field strength, E V cm^{-1}. This can be written as:

$$v \propto E \tag{1.1a}$$

or $v = \mu \times E$, where μ is a constant.

We call μ the *electrophoretic mobility* of the molecule. It is the ratio of migration velocity v to field strength E, and is the migration velocity of a particle under the influence of an applied electric field of strength 1 V cm^{-1}.

Π What units does electrophoretic mobility (μ) have?

We can work this out in the following way:

From Eq. 1.1a, $v = \mu \times E$, and $\mu = v/E$.

The units of migration velocity v are cm s^{-1}, and the units of field strength E are V cm^{-1}.

Hence

$$\mu = \frac{\text{migration velocity } (v)}{\text{field strength } (E)} = \frac{\text{cm s}^{-1}}{\text{V cm}^{-1}}$$

from which we can see that the units of μ are $cm^2\ V^{-1}\ s^{-1}$.

The sign of the electrophoretic mobility of a molecule is either + or −, depending on its net charge. A particle with no net charge will have zero electrophoretic mobility, and should not move under the influence of an applied electric field.

If we measure the migration velocities v' and v'' of a molecule under the influence of applied fields of strength E' and E'' respectively, the electrophoretic mobility (μ) of the molecule in the first situation will be v'/E' and in the second situation will be v''/E''.

From these equations we can see that $\mu = v'/E' = v''/E''$. Therefore, by measuring the migration velocity of the molecule at various field strengths, we should be able to calculate a value for μ.

Try this out for yourself by answering the following SAQ:

SAQ 1.1a

A molecule has a migration velocity of 1.60×10^{-3} cm s^{-1} under the influence of an applied field of strength 8 V cm^{-1}, and a migration velocity of 4.20×10^{-3} cm^{-1} when the applied field strength is 20 V cm^{-1}. Calculate: ⟶

SAQ 1.1a (cont.)

(*i*) the electrophoretic mobility (μ) of the molecule, and

(*ii*) its migration velocity in a field of strength 12 V cm^{-1}.

Now let us consider what will happen if an isolated molecule with net charge Q is suspended in a liquid medium to which a uniform electric field of strength E V cm^{-1} is applied. The electric field will exert a force on the particle. The magnitude of this force depends on both the field strength (E) and the size of the charge (Q) on the particle, and has a magnitude equal to $Q \times E$, as shown in Fig. 1.1b.

Under the influence of this force, the molecule will start to move towards the electrode of opposite charge. It will accelerate rapidly from rest, but this acceleration will be opposed by a frictional force as the molecule starts to move through the medium. This frictional force is called *viscous drag*:

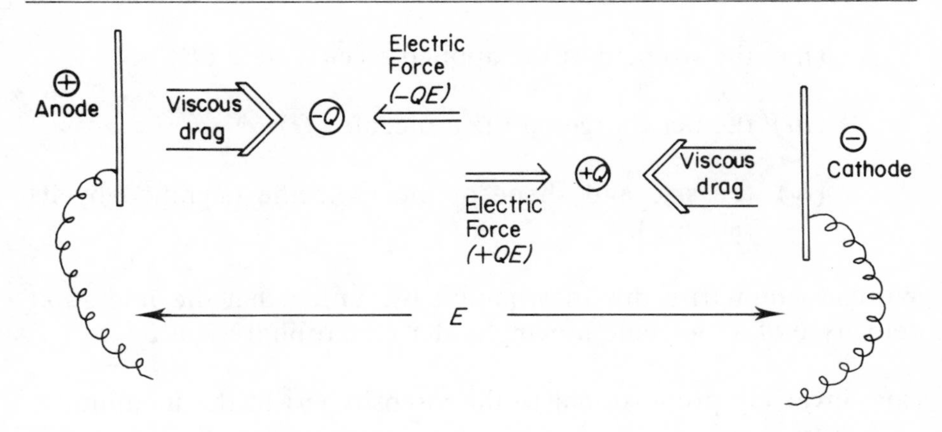

Fig. 1.1b. *Forces acting on charged molecules moving in an electric field*

The magnitude of the force of viscous drag depends on the viscosity, η, of the medium. The more viscous the medium, the more resistance will there be to movement of a molecule through it.

Viscous drag also depends on the size and shape (usually signified by the radius r) of the moving molecule. A small compact molecule will encounter less resistance to its movement than one that is large and/or irregular in shape.

When the electric force and the force of viscous drag on a molecule are exactly equal there will be no net force acting on it and therefore it will no longer accelerate. It has reached a steady state and it will continue to move at the velocity it has now reached, provided there is no change in any of the parameters of the system. Under these conditions, this velocity is actually the *migration velocity* (v) which we discussed earlier.

Π What are the parameters of an electrophoretic system that must not be changed if we want the migration velocity of a molecule in that system to remain constant?

The important parameters are:

(i) the viscosity of the medium (η)

(*ii*) the strength of the applied electric field (E)

(*iii*) the net charge on the molecule (Q)

(*iv*) the size and shape of the molecule (signified by its radius r).

We can summarise this information by stating that the migration velocity v of a molecule moving under electrophoresis is:

(*i*) inversely proportional to the viscosity (η) of the medium, ie $v \propto 1/\eta$,

(*ii*) proportional to the strength of the applied field (E), ie $v \propto E$,

(*iii*) proportional to the net charge on the molecule (Q), ie $v \propto Q$,

(*iv*) inversely proportional to the size of the molecule (r), remembering that molecular shape also has some effect, ie $v \propto 1/r$.

It follows from (*iii*) and (*iv*) that if molecules of different sizes (typified by their radius, r) but with the same net charge (Q) are moving through an electrophoresis system, their migration velocities will be inversely related to their sizes, ie large molecules will move more slowly than smaller ones with the same net charge. The electrophoretic technique has been very useful for separating molecules either on the basis of a difference in their charges or, if their charge is the same, by a difference in their molecular size.

Before we finish this Section, it is important to point out that the electrophoretic theory we have been discussing has been based on the consideration of perfectly spherical molecules moving in a fluid medium that exerts no effect, other than viscous drag, on molecules moving through it. Unfortunately, no molecules are perfectly spherical in shape, so the frictional forces opposing their movement will tend to be larger than for perfectly spherical molecules. Also, electrophoresis is carried out in ionic solutions, and there will be electrostatic interactions between any charged groups on the molecules under investigation and the ions in these solutions. The molecules will become surrounded by ions of opposite charge to their own, and this will shield them from the influence of the applied electric field. The ionic environment will be partly disrupted by the electric field itself, and by the movement of the molecule through the medium. The result will be that the molecule will move with a migration velocity less than would have been predicted.

1.2. APPLICATIONS OF ELECTROPHORESIS

Biological macromolecules, including proteins and nucleic acids, are dispersed in aqueous media to form colloidal suspensions. All the particles in a colloidal suspension possess an ionic charge because of the presence of ionisable groups in the molecules. The net charge on each molecule may be positive or negative. If an external electric field is applied to the colloidal suspension, the molecules will migrate to the anode or the cathode depending on their charge. Since this is the basis of electrophoresis, this technique has found wide application in the characterisation of biological molecules. In fact, the main applications of electrophoresis have been in the separation of biological molecules, both of low relative molecular masses such as amino acids, small peptides, and nucleotides, and of high relative molecular masses such as proteins and polynucleotides, including RNA and DNA molecules.

1.2.1. Separation of Proteins

Proteins are linear polymers of amino acids, and are found in all living systems. They are macromolecules of relative molecular mass ranging from several hundreds to many thousands. Only 20 amino acids are commonly found in plant and animal proteins, and these are combined in countless ways to form an enormous variety of different protein molecules. This tremendous diversity is needed because of the vast number of different functions that proteins perform in living systems. These 20 amino acids are listed in Table 5 at the end of the Unit.

Protein molecules consist of a 'backbone', formed from a large number of peptide bonds, from which protrude the side-chains of the individual amino acid residues. The nature of the component amino acid side-chains is responsible for the varied individual properties of different proteins.

From the point of view of electrophoresis, the most important properties of proteins are their size (ie their radius, *r*) and their net charge (Q). Every amino acid has a carboxylic acid group and an amino group, but in protein molecules these are used up in formation of the peptide bonds, with the exception of those at the N-terminus and the C-terminus of the protein, as shown below:

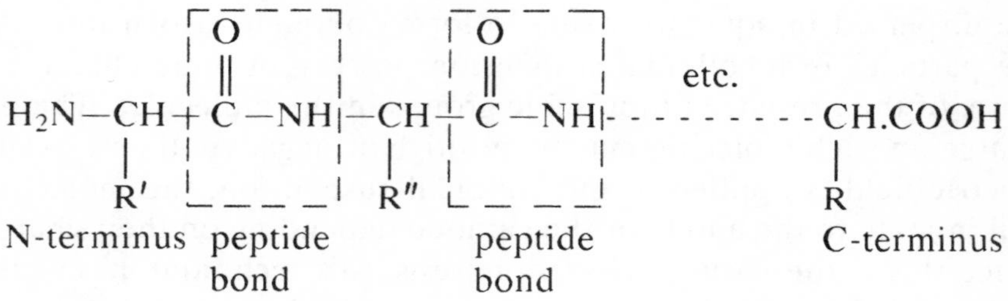

```
             ┌ O ─ ─ ┐      ┌ O ─ ─ ┐
             │ ||    │      │ ||    │      etc.
H2N—CH─┼ C —NH│—CH─┼ C —NH│- - - - - - - - CH.COOH
     |       │       │   |   │       │                |
     R'      └ ─ ─ ─ ┘   R"  └ ─ ─ ─ ┘                R
N-terminus  peptide          peptide              C-terminus
            bond             bond
```

If you look at Table 5 you will see that every amino acid has two ionisation constants (expressed as pK_a values) which represent the ionisation characteristics of the 1-amino group and 1-carboxylic acid group of the free amino acid. However, some of the amino acids have more than two pK_a values, because they have ionisable groups in their side-chains. For example, aspartic acid and glutamic acid each have a carboxylic acid group in their side chains, whereas lysine and arginine each have an amine group in their side-chains.

SAQ 1.2a

By referring to Table 5, write out a list of the amino acids that have ionisable groups in their side chains, giving the pK_a value for each of them.

SAQ 1.2a

The net charge of a protein molecule in aqueous solution depends on the ionisation characteristics of the side chains of its constituent amino acids and the pH of the solution. At low pH values, the free carboxylic acid groups will be unionized and the amine groups protonated, so that the protein will have a net positive charge:

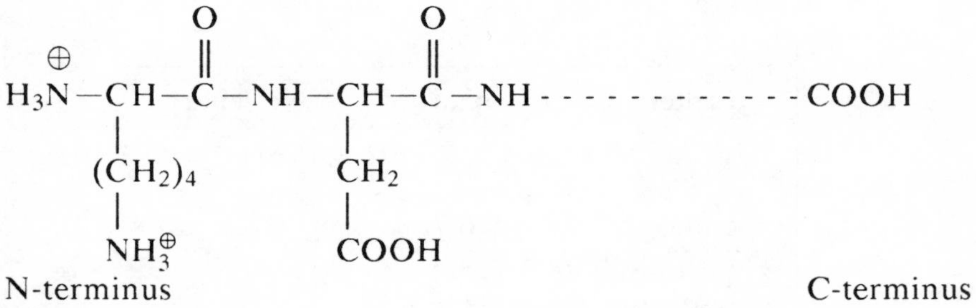

```
               O              O
 ⊕             ‖              ‖
H3N—CH—C—NH—CH—C—NH - - - - - - - - - - - - COOH
     |              |
   (CH2)4          CH2
     |              |
    NH3⊕           COOH
N-terminus                                  C-terminus
```

Conversely, at high pH values, the free carboxylic acid groups will be ionized but the amine groups will not be protonated, so the protein will have a net negative charge:

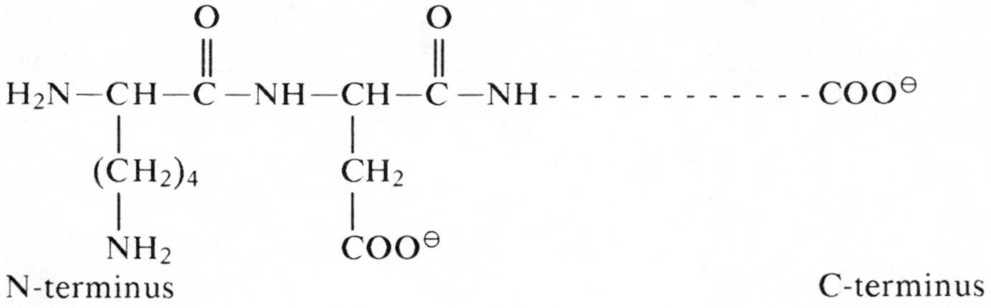

```
               O              O
               ‖              ‖
H2N—CH—C—NH—CH—C—NH - - - - - - - - - - - - COO⊖
     |              |
   (CH2)4          CH2
     |              |
    NH2            COO⊖
N-terminus                                  C-terminus
```

At intermediate values of pH, some of the free side-chain groups will be protonated and some will not, depending on their individual pK_a values. There will be one particular pH at which a protein molecule will have no net charge, because it contains an equal number of positively and negatively charged groups. This pH value is called the *isoelectric point* of the protein, and is designated 'pI'. The pI value is a characteristic of the individual protein. At pH values lower than its pI a protein will have a net positive charge, whereas at pH values higher than its pI a protein will have a net negative charge.

To give you some experience in using this concept, try answering the following SAQ:

SAQ 1.2b

The table below lists a number of proteins and their isoelectric points (pI). For each protein, decide whether its net charge will be positive or negative at (*i*) pH 3.0. (*ii*) pH 7.4. and (*iii*) pH 10.0

Protein	pKa	Net charge		
		pH 3.0	pH 7.4	pH 10.0
collagen	6.7			
serum albumin	4.8			
lysozyme	11.1			
human haemoglobin	7.1			
insulin	5.4			
cytochrome c	10.0			
pepsin	1.0			

It follows that at pH values below its isoelectric point a protein will migrate towards the cathode during electrophoresis, whereas at pH values above its isoelectric point, it will migrate towards the anode.

Π Towards which electrode will a protein molecule migrate if electrophoresis is carried out in a solution of pH exactly equal to its isoelectric point?

In this case, the protein will have no net charge, and it should therefore not migrate during electrophoresis. Electrophoresis can therefore be used to determine the pI values for individual proteins by observing the migration of the proteins during electrophoresis in a series of buffered solutions of different pH values.

A second important use of electrophoresis is suggested from the response to SAQ 1.2b. At any given pH, different proteins will have different net charges because of their characteristic pI values. They will therefore have different migration velocities under electrophoresis, and will be able to be separated from each other by this technique.

SAQ 1.2c

From the response to SAQ 1.2b, what do you suggest would be a suitable pH of solution to use for the optimum separation of human haemoglobin and cytochrome c – pH 3.0, pH 7.4, or pH 10.0? Give a reason for your answer.

SAQ 1.2c

1.2.2. Separation of Polynucleotides

In the Study Guide to this Unit, you were advised to refresh your memory about the composition and structures of the nucleic acids, RNA and DNA. It may help you at this point to look back at the responses to the preliminary SAQs 1c and 1d.

In all polynucleotides, phosphate groups form a negatively charged backbone to the molecules. Although in RNA molecules and other single-stranded polynucleotides the purine and pyrimidine bases have groups on them that can protonate in aqueous solutions, such polynucleotides do not exist as cations in aqueous solutions because they are insoluble at the low pH values that would be necessary to completely protonate their phosphate backbones. Single-stranded polynucleotides are soluble in neutral or basic solutions, and their net charge is negative because of the dissociation of the phosphate groups. Consequently, such polynucleotides will migrate towards the anode under electrophoresis.

Whereas RNA molecules consist of a single polynucleotide strand, DNA molecules (with the exception of the DNA of some viruses) consist of two polynucleotide strands with hydrogen bonds between the purine and pyrimidine bases of the adjacent strands. The resultant structure is the 'double-helix', first described by Watson and Crick in 1953. The potentially ionisable groups on the purine and pyrimidine bases are not available for ionisation in such double-stranded molecules because:

(*a*) they are protected from the solvent by being inside the double backbone of sugar-phosphate, and

(*b*) they are involved in the internal hydrogen bonding of the double helix.

DNA molecules are soluble only at basic pH values. They always have a negative charge, and will migrate towards the anode under electrophoresis.

The charge-to-size ratio is virtually constant for all single-stranded polynucleotides and for all double-stranded DNA molecules, since it depends on the ionisation of the phosphate backbone. Consequently, these molecules can be separated from each other by electrophoresis only on the basis of their sizes. This is achieved by the use of 'sieving gel' techniques, which will be described in Section 3.3. of this Unit.

The technology associated with the electrophoresis of single and double-stranded polynucleotides has revolutionised the study of RNA and DNA molecules in the fields of biochemistry and genetics. Electrophoresis is now used extensively as a tool in these areas, both for research and for routine diagnositic purposes. Two especially useful features of the techniques are:

— only very small amounts of material are needed for analysis, and

— certain characteristics of the separated molecules can be studied without eluting them from the gels.

1.2.3. Analytical and Preparative Electrophoresis

As you will see in the following Sections, electrophoresis has been widely used for the analysis of mixtures. After separation, the components of the mixture are located and identified by appropriate detection methods which will be described in Section 5.

Electrophoretic techniques have also been used on a preparative scale. Large volume samples are separated into their components and these are collected from the electrophoresis system and used for further study. Such preparative methods have been used for proteins and polynucleotides.

However, there are problems encountered in the scale-up of analytical electrophoretic techniques for their application to preparative work. These include the following:

- The heat generated during electrophoresis increases as the dimensions of the system increase.

- Loss of resolution of separated components occurs as the thickness of the electrophoretic support medium increases.

- It is difficult to recover the separated components in good yield, and with undiminished biological activity, after the electrophoresis.

Despite these and other difficulties, many different methods have been described for preparative electrophoresis, and many useful results have been obtained. Nonetheless, for the large-scale preparation of biological macromolecules, the most popular techniques are chromatographic.

Summary

The origins of electrophoretic techniques have been described, and the fundamental theory of electrophoresis presented. Applications of electrophoretic techniques to the separation and analysis of protein and polynucleotide samples have been outlined and the problems involved in scale-up of analytical electrophoresis to the preparative mode discussed.

Objectives

Having completed this Part, you should now be able to:

- give a brief outline of the principles and applications of electrophoresis;

- describe the properties of protein and polynucleotide molecules that make them suitable for separation by electrophoresis.

2. Types of Electrophoretic System

There are three different types of electrophoretic system:

— moving boundary,
— zone,
— steady state.

The original work of Tiselius in the 1930s was with *moving boundary electrophoresis*, and it is from this technique that the others have been derived. From 1950 onwards a number of zone electrophoretic techniques were developed. Steady state electrophoretic techniques are a more recent development, and are of two types – *isoelectric focussing* and *isotachophoresis*.

2.1. MOVING BOUNDARY ELECTROPHORESIS

Moving boundary electrophoresis was used quite widely between 1935 and the 1950s, and various designs of equipment for this technique became available commercially. The principal use was in protein research, where it provided valuable information. The equipment used consists of a U-shaped cell with a rectangular cross-section as shown in Fig. 2.1a:

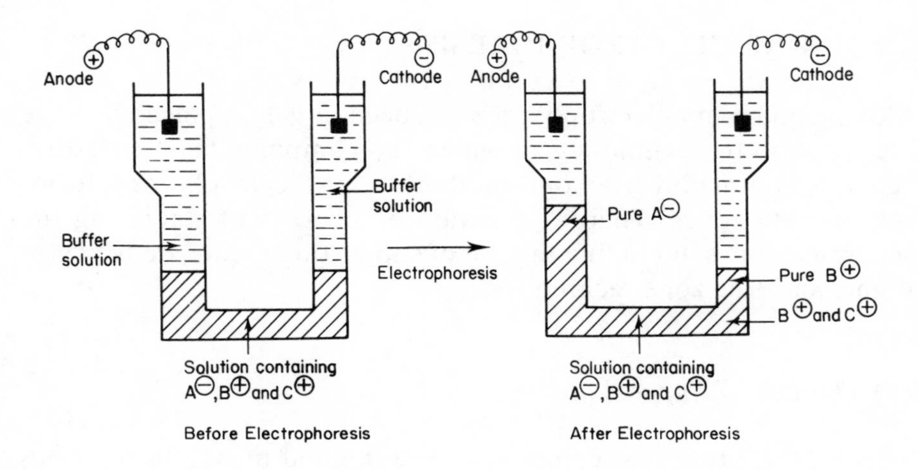

Fig. 2.1a. *Moving boundary electrophoresis*

The equipment is partially filled with the protein solution of interest, and over it is layered a buffer solution. Electrodes are immersed in the buffer solution, and the entire system is placed in a thermostatted bath. When the voltage is applied, charged molecules migrate towards the appropriate electrodes. If several components are present in the sample, and if their electrophoretic mobilities differ, their migration can be observed as multiple moving boundaries in the system. Since the protein solution has a refractive index a little higher than that of the buffer, there will be a local change in refractive index at the protein boundary in the U-tube and this can be detected by optical methods, usually using 'Schleiren' optics.

After electrophoresis is complete, fractions can be obtained that contain separated components of the original sample and these can be analysed chemically and biologically, as appropriate. In this way, biologically active substances present in concentrations too low to be detected by optical means may be determined.

A major drawback of moving boundary electrophoresis is that only the slowest and most rapidly moving components of a mixture can be obtained in pure form. However, it has been a useful method for the determination of the complexity of heterogeneous samples.

2.2. ZONE ELECTROPHORESIS

Moving boundary electrophoresis has been largely replaced by zone electrophoretic techniques, in which the components of a mixture separate completely from one another during electrophoresis, forming discrete zones. There are a number of ways of stabilising the separated zones, including the use of supporting media, density gradients and free zone techniques.

(*a*) *The Use of Support Media*

This is by far the most commonly used method of stabilising zones of electrophoretically separated mixtures. A porous medium, such as filter paper, cellulose acetate film, or a gel, is used to provide a support through which the sample under investigation moves during electrophoresis. The sample is applied to the support medium as a spot or a narrow band, at a point called the *origin*. A typical apparatus is shown in Fig. 2.2a:

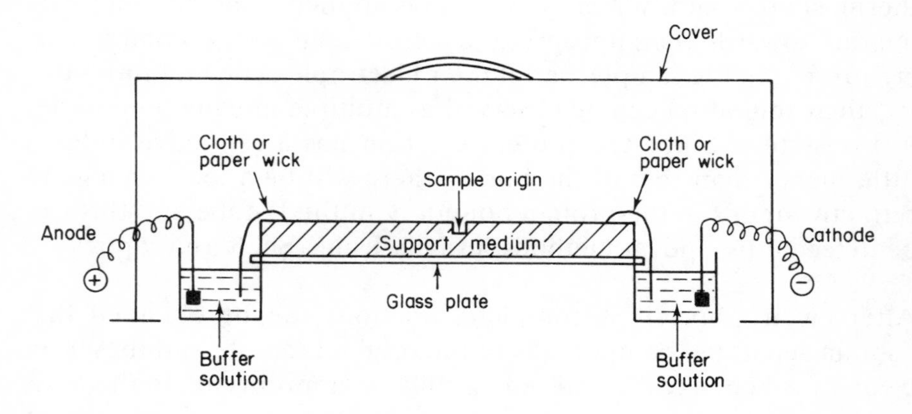

Fig. 2.2a. *Apparatus for horizontal electrophoresis*

During the run, the components of the sample separate into bands which are kept distinct by the presence of the support medium.

SAQ 2.2a

> Make a list of the properties that you think would be desirable in a substance which is to be used as a support material in zone electrophoresis.

In Part 3 we will discuss in detail the different types of support medium used in zone electrophoresis.

(*b*) *Density Gradient Stabilisation*

The basis of this technique is to carry out the electrophoresis in a density gradient column, which prevents convection currents disturbing the separated zones. Gradients are usually made up of solutions of substances such as sucrose, glycerol, ethylene glycol, as these are very soluble in aqueous solutions, do not ionise, and are unlikely to interact with the material to be separated. The density gradient is a concentration gradient of one of these solutes.

Π Why do you think that it is important that the density gradient solute does not ionise?

The solute forming the density gradient should not have any charge under the conditions of the electrophoresis. If it did, the solute molecules would move towards the appropriate electrode, and also could interact with the sample molecules and interfere with their migration.

(*c*) *Free Zone Electrophoresis*

This rather complex technique involves either rotation of the electrophoresis vessel during the run, or the streaming of a continuous film of buffer solution across the electrophoretic system, in a direction perpendicular to the electric field.

These two methods have the advantage that it is relatively simple to recover the separated components at the end of the analysis. However, they have not been extensively used, mainly because the equipment required is quite elaborate, and consequently expensive.

2.3. STEADY STATE ELECTROPHORESIS

The characteristic of this method is that after electrophoresis has proceeded for a certain length of time, a steady state is attained in which the widths and positions of the zones of the separated components do not change with time.

2.3.1. Isoelectric Focussing

There are several good books and review articles about this technique – for example, that of Righetti and Drysdale (1976). During isoelectric focussing, a stable pH gradient is created between the anode and the cathode. Under the influence of an applied electric field, the charged molecules in the sample move through the medium until they eventually reach a position in the pH gradient where their net charge is zero and they will migrate no further. For polypeptides and proteins, this pH is their *isoelectric point*. Molecules with the same isoelectric point (pI) become concentrated in a narrow band at that pH value, as shown in Fig. 2.3a. The individual pI bands can be collected from the apparatus, and examined chemically and biologically as appropriate.

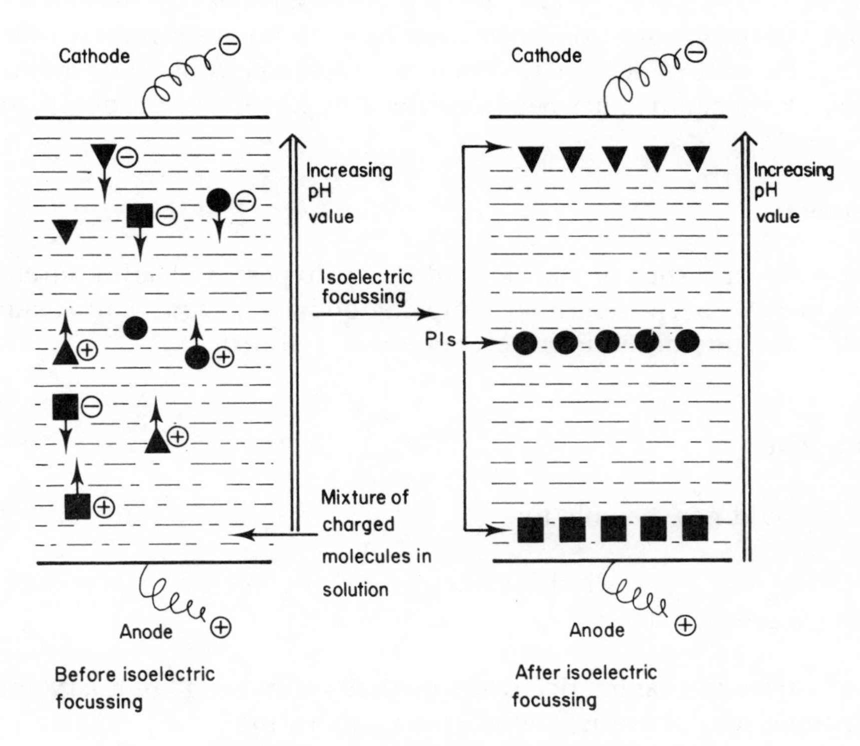

Fig. 2.3a. *Isoelectric focussing*

Isoelectric focussing is a technique of very high resolving power, and it can be used preparatively. One disadvantage is that some proteins precipitate at their isoelectric point, because they have no charge, and this can make their recovery difficult.

2.3.2. Isotachophoresis

Although the principles of isotachophoresis were first discovered in 1897 by Kohlrausch, it is only since the early 1970s that it has been used as an analytical technique. It is particularly useful for the separation of charged molecules of small relative molecular mass, such as certain drugs and polypeptides of medical interest.

In isotachophoresis, all the charged components become *stacked* one behind the other according to their electrophoretic mobilities. To separate two ions from each other by isotachophoresis there must be a difference of at least 10% between their electrophoretic mobilities. This technique has been described by Karol and Karol (1978).

Summary

The characteristics of the three different types of electrophoretic system have been described – moving boundary, zone and steady state – and the differences between them outlined.

Objectives

You should now be able to:

- distinguish between the principles of the three types of electrophoresis system;
- illustrate diagrammatically the methods of moving boundary and zone electrophoresis, and isoelectric focussing;
- discuss the criteria used in selecting an appropriate type of electrophoresis system for the separation of molecular components in a mixture.

3. Support Media used in Zone Electrophoresis

This section will examine in some detail the characteristics of the various support media that have been used in zone electrophoresis systems. We will mention the advantages and disadvantages of each type of medium, and the applications for which they have been shown to be particularly useful.

3.1. FILTER PAPER

The use of filter paper as a support medium was introduced in the early 1950s and derived directly from the use of paper in chromatography. Paper electrophoresis found wide application because it was simple and quick and required relatively simple apparatus. Fig. 3.1a is a diagram of a typical apparatus used for vertical electrophoresis with paper. Other systems have the paper lying horizontally.

In paper electrophoresis, the paper is moistened then placed on a supporting rack. The samples are applied with a capillary applicator as spots or streaks. The paper is put into the apparatus with each of its ends dipping into an electrode vessel containing electrophoresis

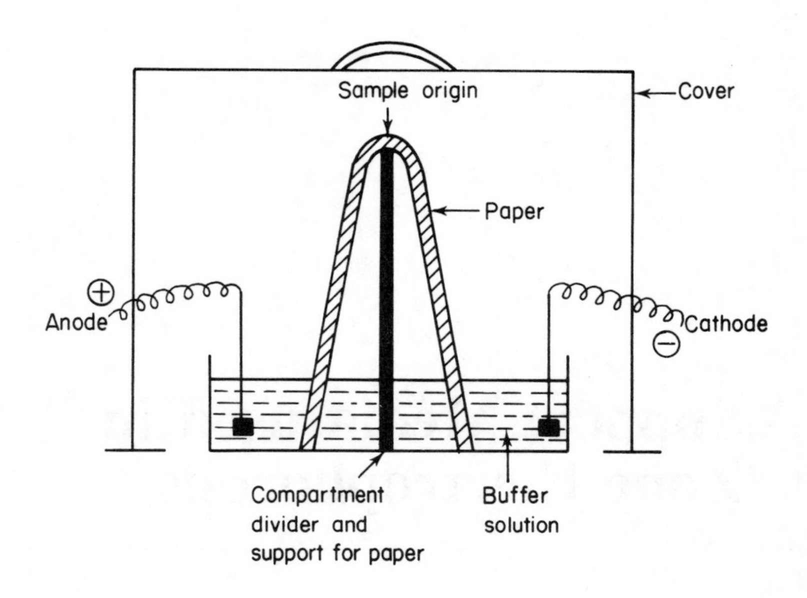

Fig. 3.1a. *Apparatus for vertical paper electrophoresis*

buffer, and a lid is put on to stabilise the humidity and prevent the paper drying out. In some types of apparatus there is a system for keeping the paper cooled during the run; otherwise, the apparatus can be set up in a refrigerated room (cold room).

At the end of the run the paper is carefully taken out and dried. It can then be examined by appropriate detection techniques, as described in Section 5. With some detection methods (eg staining) the separated components can be located directly on the paper, which can then be retained as a semi-permanent record of the results.

Electrophoresis on paper has been used to separate charged molecules of many types, including proteins, oligonucleotides, carbohydrates and lipids. However, there are some disadvantages with paper electrophoresis.

(*i*) If the paper is too thin it will tear easily, especially when wet; if it is too thick the boundaries between the zones become distorted.

(*ii*) Many biological molecules, particularly proteins, interact with hydroxyl groups on the cellulose in paper, so that the migration of the molecules is hindered. This can lead to 'tailing' of individual protein zones. Since different proteins interact differently with the cellulose, this can alter the separation characteristics of protein components of a mixture.

(*iii*) The quality of the paper is important. It must contain at least 96% α-cellulose, since the capacity of the paper to absorb water and retain its electrical conductivity depend upon this.

(*iv*) Another significant problem is the phenomenon known as *electro-osmosis*. Paper will take on a negative charge when in contact with water, probably because of the presence of very small numbers of carboxylic acid groups in the cellulose matrix. Since the cellulose is fixed in position during electrophoresis, the hydronium ions resulting from dissociation of the carboxylic acid groups (H_3O^+) move instead (towards the cathode) when an electric field is applied to the system. Consequently, the transport of molecules in paper electrophoresis is the sum of their electrophoretic movement and the anode-to-cathode electro-osmotic effect.

A correction factor must be applied for the electro-osmotic effect to get a true value for electrophoretic mobility. This can be done by applying along with the sample some coloured substance which does not ionise. Blue dextran, available commercially, has been recommended for this purpose. Electro-osmosis is measured as the distance between the sample origin and the position of the blue dextran zone after electrophoresis.

The field strength used in paper electrophoresis is usually low, around 20 V cm^{-1}. Such a low voltage is inefficient for the separation of small molecules such as nucleotides and amino acids, because their small size allows considerable spreading of the zones by diffusion during the time of the run. A modification of the standard paper electrophoretic technique has been developed for molecules of small relative molecular mass, in that the speed of separation of the molecules is increased and their diffusion minimised by using a field strength of around 200 V cm^{-1}. The *high voltage* leads to considerable heating of the paper so that an efficient cooling system is necessary.

Paper electrophoresis has been adapted, by the use of thick papers, for the preparative electrophoresis of proteins and oligonucleotides. After the run, the separated zones of components can be eluted from the paper with an appropriate buffer. However, the resolution of separated components is poor.

SAQ 3.1a

To summarise what we have been discussing in this section, try writing out two lists, one of the advantages and the other of the disadvantages of paper as a medium for zone electrophoresis.

3.2. CELLULOSE ACETATE

Some of the problems associated with paper electrophoresis were overcome by the introduction of the use of cellulose acetate membranes (Kohn, 1957). In cellulose acetate most of the hydroxyl groups of cellulose have been esterified by acetylation:

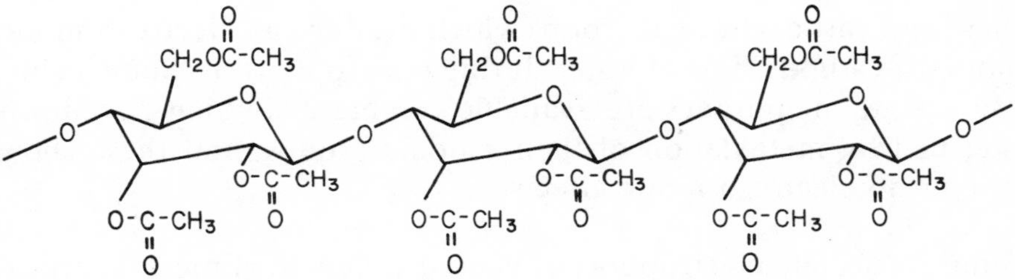

Fig. 3.2a. *Cellulose acetate*

Consequently, there are few, if any, hydroxyl groups available to interact with sample molecules. The problems of adsorption of samples onto the cellulose and the subsequent distortion of their separation are therefore avoided, and electrophoretic separations are achieved more quickly with cellulose acetate than with paper. Furthermore, cellulose acetate membranes are more homogeneous in structure than paper is.

Cellulose acetate can be made transparent simply by impregnating it with mineral oil of refractive index equal to that of the membrane, which is an advantage in the spectrophotometric determination of separated components. Moreover, if staining is used as a detection method there is little background staining with cellulose acetate.

Cellulose acetate dissolves easily in various solvents, allowing simple recovery of separated components from the membrane after electrophoresis. However, care must be taken to select a solvent that will not damage the components of interest.

In practice, electrophoresis on cellulose acetate membranes is similar to paper electrophoresis. Usually a closed horizontal arrangement is used. Care must be taken that the membranes do not dry out, which is indicated by the appearance of white spots on the membrane surface.

After electrophoresis, the cellulose acetate is dried in an oven at 80–100 °C, before being examined by appropriate detection methods.

A disadvantage of cellulose acetate is that it may contain some sulphate and carboxylic acid groups which produce an electro-osmotic anode-to-cathode flow of water during electrophoresis. Some cellulose acetate membranes are available commercially that have been esterified by methylation of their anionic groups, and these show little or no electro-osmotic effects.

Cellulose acetate electrophoresis is used widely in clinical laboratories for the separation and analysis of proteins in biological fluids. It can be used successfully with small amounts of sample (in the microlitre range) and the run time is short (sometimes less than an hour). For such small samples the fact that cellulose acetate is an expensive material is relatively unimportant.

3.3. GEL MEDIA

As has been indicated, there are several problems associated with the use of both paper and cellulose acetate as support media in electrophoresis. The introduction of gels as an alternative overcame some of these problems and was developed from their use in chromatography. A gel is a three-dimensional polymeric network with a random structure. The gels used in electrophoresis are cross-linked polymers that should be inert and not interact with the molecules under study. It is particularly important that the material does not have any ionisable groups.

Unlike those used in chromatography, the gels used for electrophoresis are not granular. Instead, they are formed into continuous columns or slabs, which gives them very different properties compared to chromatography gels. An electrophoresis gel consists of a network of polymer molecules surrounded and penetrated by buffer. The spaces between the gel molecules are the *pores* of the gel. When an external voltage is applied, charged sample molecules migrate towards the appropriate electrode. A molecule moving through a continuous gel experiences a frictional resistance to its movement which is related to the relative sizes of the gel pores and the radius of the molecule. A molecule smaller than the average pore size of the gel will be able to move relatively easily through the gel, whereas a molecule larger than the average pore size will be held back by a resistance that depends on its radius. If a molecule is very much larger than the gel pores it may not be able to enter the gel at all. A continuous gel therefore exerts a *sieving* effect on the passage of molecules through it.

With gels such as those used as support media in electrophoresis (eg starch, polyacrylamide, agarose) the average pore size depends on the concentration of polymer. The higher the concentration of polymer the more dense the gel network and the smaller the average pore size. If we know the approximate dimensions of the molecules in our sample, we can tailor the pore size of the gel to suit them. It is this flexibility in design of the characteristics of the support medium that makes gel electrophoresis such a useful technique in analytical chemistry.

Fig. 3.3a illustrates diagrammatically the separation of a mixture of molecules of different sizes by electrophoresis through a sieving gel.

For molecules with the same net charge, smaller ones will migrate faster than larger ones. You should recall that in electrophoresis the rate of migration of a molecule depends on the magnitude of its net charge. A large molecule with a high net charge could move with the same velocity as a small molecule with a low net charge, so that they would appear in the same zone at the end of the run. Therefore, if a sample gives only one band in gel electrophoresis this does not necessarily mean that there is only one component in that band.

Π Suggest how you could verify the purity (homogeneity) of a protein sample using gel electrophoresis.

To verify the homogeneity of a sample by gel electrophoresis, analysis should be carried out using:

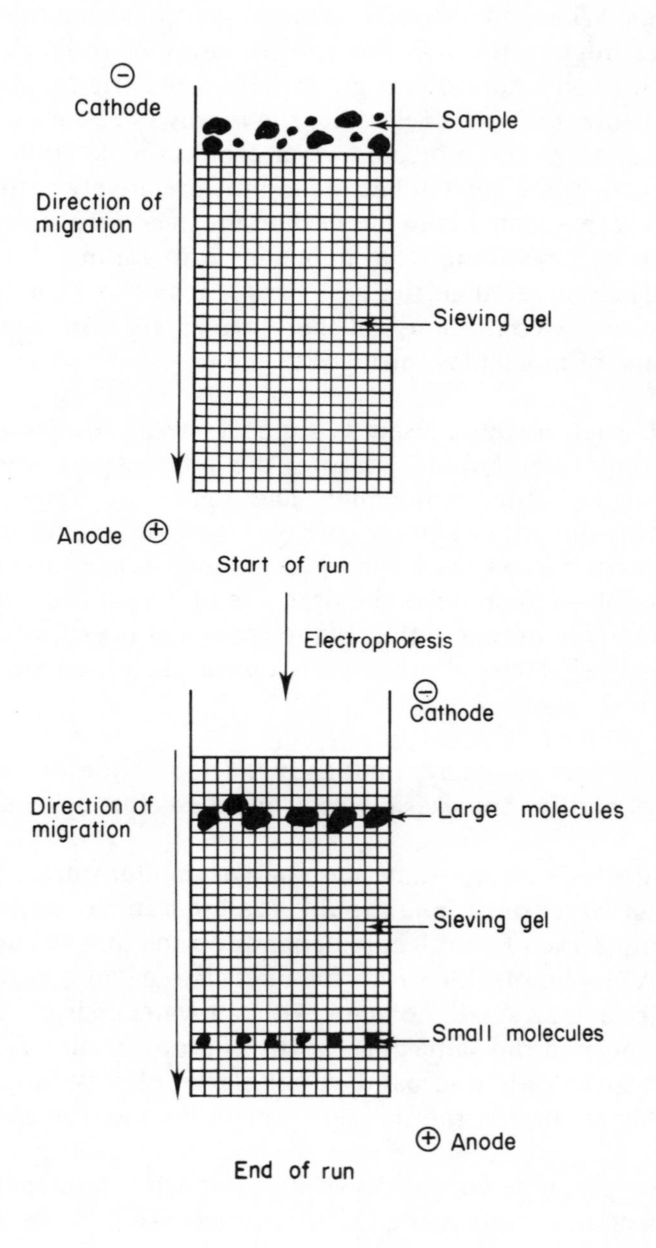

Fig. 3.3a. *Separation of molecules by electrophoresis in a sieving gel*

(*i*) several different concentrations of gel (ie with different pore sizes)

(*ii*) buffers of different pH values, to change the net charges of the molecules.

3.3.1. Starch Gels

The earliest work in the gel electrophoresis of proteins used starch as the support gel. Suitable starch gels can be prepared from hydrolysed potato starch, which is heated in electrophoresis buffer with constant swirling until the suspension becomes transparent (Poulik and Smithies, 1958), and is then poured into a mould.

The usual equipment for starch gel electrophoresis has the gel mounted horizontally, as shown in Fig. 2.2a, although for some purposes a similar but vertical arrangement has been used. Contact between the electrode buffer and the gel is made by paper or cloth wicks. Samples are absorbed onto small pieces of filter paper, which are then placed into slots cut in the gel which is then covered with cling-film to prevent it drying out. After electrophoresis, the sample papers are removed, and the gel is sliced into several layers using a fine wire. This provides several *duplicate* gels which can be examined by different detection methods.

Starch gel electrophoresis is a cheap, easy and reasonably quick analytical technique. Since the diffusion of proteins is low, and since several different detection methods can be used on duplicate gel slices, the technique has been widely used for the analysis of proteins, particularly isoenzymes (different protein molecules that have the same enzyme activity).

However, the pore size of starch gels can vary within only a narrow range since it is determined by the starch concentration, which cannot be changed much without the gel becoming either too soft or too stiff.

Furthermore, starch contains some negatively charged side-chains which may interact with protein molecules and hinder their migration. These can also lead to an electro-osmotic effect during electrophoresis (see Section 3.1).

3.3.2. Polyacrylamide Gels

Although a starch gel often remains the medium of choice for the electrophoretic analysis of isoenzymes, it has been superceded in many laboratories by polyacrylamide gels.

The main advantages of polyacrylamide compared to starch are:

(*i*) the concentration of polyacrylamide can be varied within wider limits than the concentration of starch, without making the gels unmanageable. Since the average pore size of polyacrylamide gels decreases with increasing concentration of acrylamide, the amount of molecular sieving can be finely controlled.

(*ii*) the adsorption of proteins by polyacrylamide is negligible

Polyacrylamide gels are prepared by cross-linking acrylamide with N,N′-methylenebisacrylamide. The reaction requires an initiator and a catalyst (cross-linking reagent). Commonly used initiators are ammonium or potassium persulphate (for chemical polymerisation) and light and riboflavin (for photopolymerisation). The most commonly used cross-linking reagent is N,N,N′,N′,-tetramethylethylenediamine (TEMED). The reaction takes place via vinyl polymerisation, and gives a randomly coiled gel structure:

Acrylamide: $CH_2{=}CH{-}C({=}O){-}NH_2$

N,N′-methylene bisacrylamide: $CH_2{=}CH{-}C({=}O){-}NH{-}CH_2{-}NH{-}C({=}O){-}CH{=}CH_2$

Acrylamide + N,N′-methylene bisacrylamide $\xrightarrow{\text{catalyst + initiator}}$ polyacrylamide:

$-CH_2-CH(CONH_2)-CH_2-CH(CONH_2)-CH_2-CH(C{=}O)-CH_2-$

$\qquad\qquad\qquad\qquad\qquad\qquad\qquad\qquad\quad |\ NH-CH_2-NH$ (Cross Link)

$-CH_2-CH(CONH_2)-CH_2-CH(CONH_2)-CH_2-CH(C{=}O)-CH_2$

polyacrylamide

Since both types of polymerisation reaction are very sensitive to certain chemical impurities that may be present in the components of the gel mixture, it is important to use very high quality reagents in the preparation of polyacrylamide gels. Both the acrylamide and the N,N′-methylenebisacrylamide should be *white* crystalline substances.

When making polyacrylamide gels, the ratio of cross-linking agent to acrylamide is a critical factor as it determines the pore size of the gel and influences its mechanical properties.

Polyacrylamide gels may be made as *columns* or *slabs*. In the first case, the gel is polymerised in a glass or plastic cylindrical tube of internal diameter about 5 mm and length 70–100 mm. Longer tubes can be used to separate components of similar mobilities, but they take several hours to run, which allows time for considerable diffusion of the separated zones. During polymerisation, the gel is gently overlayered with water to ensure that the gel surface is flat. After polymerisation, the water layer is removed, the gel surface rinsed with buffer, and the sample gently layered on top of the gel, usually in a concentrated solution of sucrose to increase its density. The tube is filled with buffer and placed vertically into a supporting rack in the electrophoresis apparatus. Cooled buffer is placed in each electrode vessel, care being taken to ensure that no air bubbles are trapped at the top or bottom of the gels. A small amount of ionisable tracking dye is added to the top electrode vessel, so that the progress of the electrophoretic run can be monitored. Bromophenol blue is the usual anionic dye, and methylene blue is a suitable cationic dye.

One arrangement for tube gel electrophoresis is shown in Fig. 3.3b.

The electrophoresis is usually started with a current of 1 mA per tube for about half-an-hour to allow the samples to enter the gels, after which it can be increased to 2–5 mA per tube. It is advisable to keep the system cool, if possible, say in a refrigerated room, and also to circulate the electrophoresis buffer (so that pH changes are avoided).

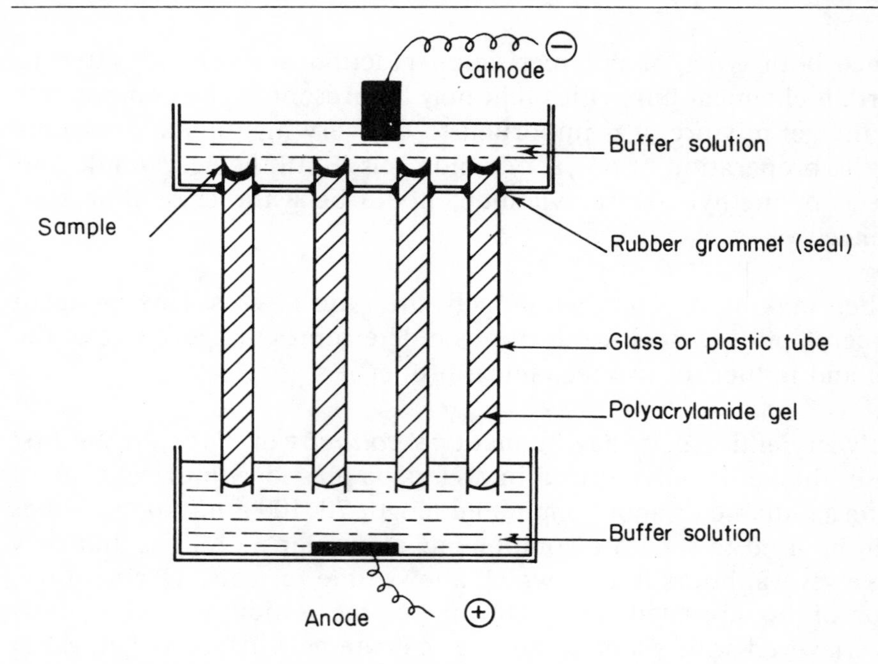

Fig. 3.3b. *Apparatus for tube gel electrophoresis*

After the tracking dye has migrated almost to the bottom of the gels, the power is turned off and the gels removed from their tubes. Proteins can be 'fixed' in place by precipitating them in the gels with a solution of trichloroethanoic acid. They can then be stained or assayed for radioactivity, as appropriate.

Polyacrylamide slab gels are useful for analysing several samples under the same electrophoretic conditions. They have other advantages compared to tube gels:

(*i*) their heat dissipation is better,

(*ii*) they can be used either horizontally or vertically. The horizontal arrangement is used when the gel concentration is low, and the gel is soft,

(*iii*) they have higher resolving power.

Slab gels are prepared in essentially the same way as tube gels. The apparatus used for horizontal electrophoresis is similar to that described earlier (Fig. 2.2a). The vertical apparatus is illustrated in Fig. 3.3c:

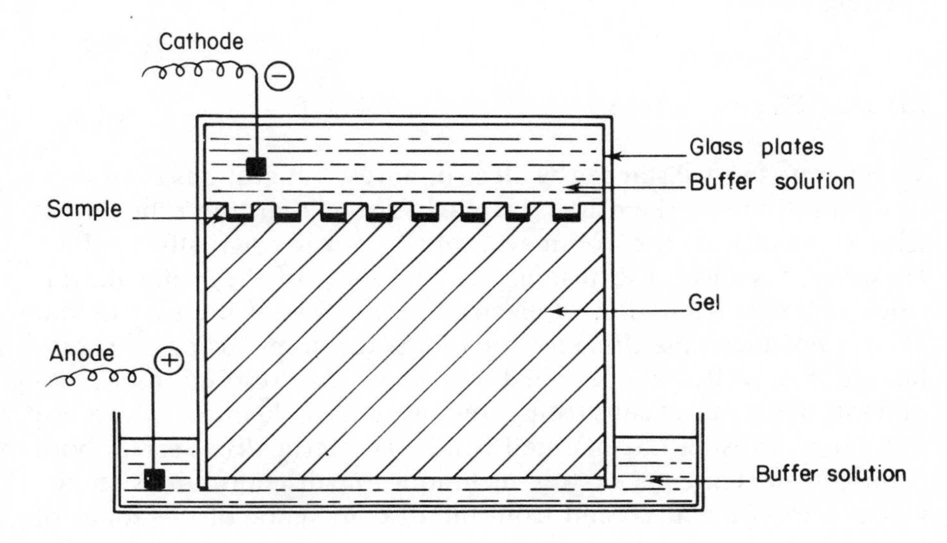

Fig. 3.3c. *Apparatus for vertical slab gel electrophoresis*

A slot former, or *comb*, is inserted into the apparatus before the gel polymerises and is removed after polymerisation, leaving sample wells separated from each other by continuous strips of gel. The samples are loaded into the wells usually in buffer containing glycerol or sucrose to increase their density.

The apparatus used for slab gels has until recently been considerably more expensive than that used for tube gels, and slab gels need more acrylamide. Consequently, they are also more expensive to use than tube gels. However, their use is increasing for the analysis of proteins, small RNA molecules, and fragments of DNA. Since single-stranded and double-stranded polynucleotides have constant charge-to-size ratios, their electrophoretic separation is achieved on the basis of their sizes (see Section 1.2.2). The sieving gel properties of polyacrylamide gels, particularly slab gels, have therefore been widely used for the analysis of mixtures of these molecules.

The usefulness of polyacrylamide as a support medium for electrophoresis has led to the development of many variations of the original technique, specifically designed for particular purposes. Three of these will be discussed in some detail in the following Sections:

(*a*) *Disc Electrophoresis*

To separate two substances by electrophoresis, migrations must continue until one of them has travelled further than the other by at least the width of the volume in which it was originally applied. However, because of diffusion, the sharpness of the zones diminishes with increasing time of electrophoresis. In 1964, Ornstein and Davis introduced the *disc electrophoresis* technique using polyacrylamide gels, which was designed to achieve high resolution of components in very brief runs (sometimes as short as 20 minutes), and so minimise diffusion of separated zones. The term *disc* derives both from the dependence of this technique on discontinuities in the electrophoretic matrix, and from the discoid shape of the zones of the separated molecules.

Analysis of even very dilute samples becomes routine using disc electrophoresis because the various substances are automatically concentrated at the beginning of the run, just prior to their separation. The theory of the technique is rather complex, but is explained quite clearly in Ornstein and Davis's papers (Davis (1964) and Ornstein (1964)) which you could obtain in your library. Essentially, the system is composed of two different polyacrylamide gels layered one on top of the other. The lower gel is the *separating* gel (also called the *running* gel or the *sieving* gel), and its function is to allow molecular sieving of the migrating substances. In contrast, the upper gel (called the *stacking* or 'spacer' gel) is a large pore gel which has little or no molecular sieving effect. The acrylamide concentration of the stacking gel is usually 2.5% *w*/*v*, which is considerably less than that of the separating gel (usually 15% *w*/*v*). The gels are made up in buffers that differ in ionic strength and pH, both the pH and the ionic strength of the stacking gel being lower than those of the separating gel.

The basic equipment is illustrated in Fig. 3.3d:

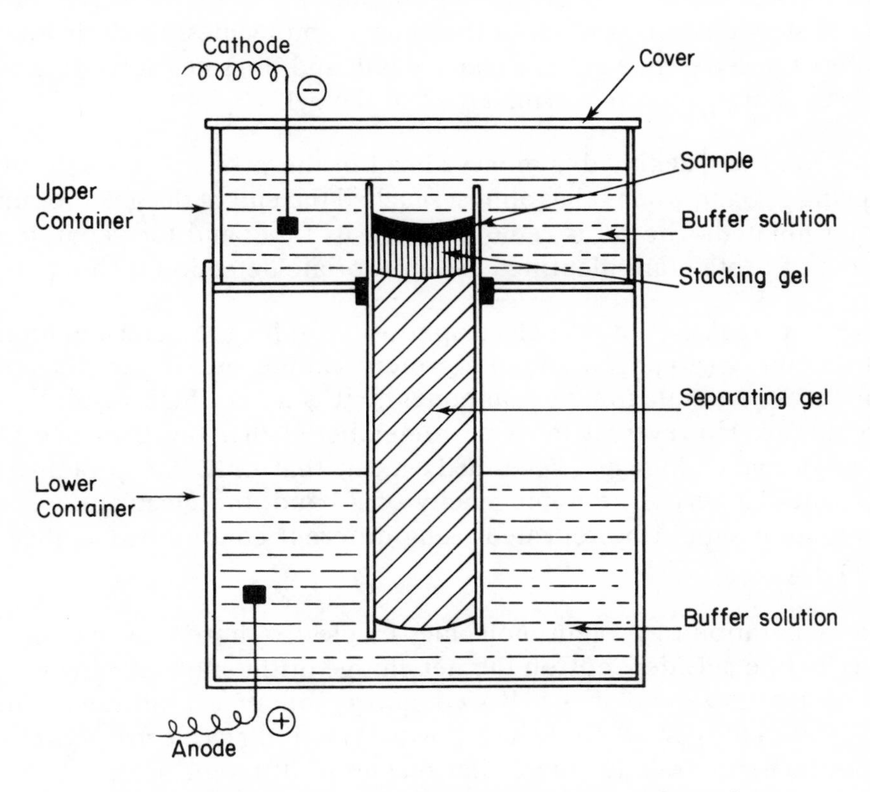

Fig. 3.3d. *Apparatus for disc electrophoresis*

The lower gel is made first in the glass column to be used for the electrophoresis. When it has set, the upper gel mixture is layered on top and allowed to polymerise. The tube is filled with electrophoresis buffer, placed in the electrophoresis equipment, and the sample in dense sucrose solution carefully layered on top of the gel. The electrode reservoirs are filled with electrophoresis buffer, a tracking dye is added to the top chamber, and the power is switched on.

During the run, the tracking dye enters the top gel and overtakes the protein molecules migrating through it. The proteins concentrate behind the dye and their concentration is completed at the lower end of the stacking gel where the increase in ionic strength (see Section 4.2.3) and the decrease in pore size of the separating gel both act to slow down migration of the charged molecules. Each protein enters the separating gel as a sharp band, and migrates according to its net charge and the sieving effect of the gel.

The band of tracking dye moves ahead of the proteins through the sieving gel, and when it has almost reached the end of the gel, the run is terminated. The gel is removed from its tube, and the separated zones of protein are identified by appropriate detection techniques.

The principal uses of disc electrophoresis are for the determination of the purity (homogeneity) of a protein sample, and for analysis of mixtures containing many components. It is a very high resolution technique. However, it must be remembered that the presence of a single band does not necessarily mean that only one protein is present. To verify the homogeneity of a band, the electrophoretic separation should be repeated using different conditions (see Section 4).

The separation of protein molecules by electrophoresis using polyacrylamide gels depends on the net charge of the molecules as well as on their size and shape. If we could eliminate the difference in their net charges we could use polyacrylamide gel electrophoresis to estimate the relative molecular masses of different proteins.

There is a relation between the rate of migration of molecules through sieving gels and their relative molecular mass, provided their net charges are the same. The smaller the gel pore size (ie the more concentrated the gel) the more slowly a molecule migrates through it. This relation is described by the equation:

$$D = a - b.\log_{10}M \qquad (3.1)$$

where D is the distance migrated by a molecule of relative molecular mass M, and a and b are constants characteristic of the electrophoresis system.

Eq. 3.1 holds well for polyacrylamide gel electrophoresis, and can be used to calculate the relative molecular mass of a molecule, as illustrated by the following SAQ:

SAQ 3.3a

The proteins myoglobin and ovalbumin have relative molecular masses (M_r), 17,200 and 43,000 respectively. During electrophoresis through a polyacrylamide gel, they migrate 1.50 cm and 5.50 cm respectively. What is the M_r of the protein, α-chymotrypsinogen, that migrates 3.25 cm in the same gel?

Since the results of this type of experiment depend on protein shape and not just on relative molecular mass, many experiments have to be carried out to ensure that the values obtained are reliable. It is not a very satisfactory way of determining the relative molecular mass of proteins – the estimate can be only approximate.

(*b*) *SDS-polyacrylamide Gel Electrophoresis*

For protein molecules of different net charges, the difference in charge can be eliminated by complexing them with the anionic detergent *sodium dodecyl sulphate* (SDS):

$$CH_3{-}(CH_2)_{10}{-}CH_2OSO_3^-{-}Na^+$$

This is the basis of a polyacrylamide gel electrophoresis method devised by Maizel (1966) and developed further by Weber and Osborn (1969).

At neutral pH, in the presence of 1% *w/v* SDS and 0.1 M 2-mercaptoethanol, most proteins dissociate into unfolded individual polypeptide chains because the 2-mercaptoethanol destroys disulphide bridges and the SDS binds to the polypeptide chains through hydrophobic interactions. Protein–SDS complexes are formed which have a random-coil conformation. The sodium ions serve as counterions to the SDS-protein complexes.

Proteins treated in this way all have an identical charge-to-size ratio because the amount of SDS bound per unit mass of protein is constant, at 1.4g of SDS per gram of protein. When subjected to electrophoresis on polyacrylamide gels containing SDS, such polypeptide–SDS complexes migrate with a velocity that is related only to their size.

The polyacrylamide gels and the electrophoresis buffer used for electrophoresis of proteins treated with SDS each contain 0.1% *w/v* SDS. Fortunately, SDS does not interfere with the polymerisation of polyacrylamide gels.

SAQ 3.3b

Give *three* reasons why the rate of migration of a polypeptide–SDS complex in SDS–polyacrylamide gel electrophoresis depends on its size.

SAQ 3.3b

If several polypeptides of known relative molecular mass are subjected to SDS–polyacrylamide gel electrophoresis and their rates of migration measured, these can be used to determine the relative molecular masses of other polypeptides analysed in the same way. This is probably the method most widely used for estimating the relative molecular masses of polypeptides and proteins.

The SDS–polyacrylamide gel technique is also useful for determining whether a protein is made up of *subunits*. If a protein has subunit structure (i.e. 'quaternary' structure) it will exhibit a different electrophoretic pattern depending on whether or not 2-mercaptoethanol was present when the protein was treated with SDS before electrophoresis. In the presence of 2-mercaptoethanol, any disulphide bonds involved in holding protein subunits together will be broken, and the protein will dissociate into its subunits. Consequently, the relative molecular mass of the SDS–protein complex will be smaller than it would have been had 2-mercaptoethanol not been used.

3.3.3. Agarose Gels

The use of agarose gel as a support medium in electrophoresis was introduced in the early 1970s. Agarose is a linear polymer of D-galactose and 3,6-anhydrogalactose which contains about 0.04% sulphate ions. It is obtained from various seaweeds. Agarose dissolves in boiling water, and when cooled to about 38 °C forms a gel that is held together by hydrogen bonds. The concentration of agarose determines the pore size of the gel – the higher the concentration the smaller the average pore size. The pores are relatively large compared to polyacrylamide gels, which makes agarose a suitable support medium for the analysis of proteins or nucleic acids that are too large to enter polyacrylamide gels (see Fig. 3.3a).

Agarose is used as slab gels for both vertical and horizontal electrophoresis. 0.8% *w/v* agarose makes a gel that is rigid enough to be held in a vertical position yet still allow molecules of relative molecular mass as high as 50×10^6 to enter it. If the horizontal arrangement is used, concentrations as low as 0.2% *w/v* agarose can be made into gels, and molecules with a relative molecular mass of 150×10^6 will migrate through it.

The equipment used for agarose gel electrophoresis is virtually the same as those described earlier for starch and polyacrylamide slab gels (Figs. 2.2a and 3.3c).

After electrophoresis, separated components in agarose gels can be detected in a variety of ways either by examining the gel directly or after drying it to a thin homogeneous film (by degassing it and warming it at 37 °C overnight).

The main problem that has been encountered with the use of agarose for gel electrophoresis is that it usually contains charged groups, principally sulphate and carboxylic acid groups. These can interact with charged groups on the ionised macromolecules under study, especially proteins, and hinder their electrophoretic migration. Furthermore, the presence of anionic groups on the support medium leads to an electro-osmotic effect, as discussed earlier (Section 3.1), which alters the electrophoretic mobilities of migrating sample molecules. These problems are minimised by using agarose that has been pretreated with alkali, which hydrolyses the anionic groups (Laas, 1972).

The physical properties of agarose gels, particularly their viscosity, are very sensitive to changes in temperature, therefore heat generated in them during electrophoresis must be removed by an efficient cooling system.

The techniques associated with agarose gel electrophoresis have revolutionised the study of DNA, and have allowed many important discoveries to be made, especially in molecular genetics. The resolution of DNA molecules by agarose gel electrophoresis is truly extraordinary.

3.3.4. Polyacrylamide–Agarose Gels

An alternative gel support medium sometimes used for electrophoresis with both tube and slab gels is one containing both polyacrylamide and agarose. The dilute solutions of acrylamide that would be required to give polyacrylamide gels of relatively large average pore size would result in gels that are very soft and difficult to handle. The inclusion of agarose in the mixture used to make the gel results in a combined polyacrylamide–agarose gel which has a relatively large average pore size and good mechanical strength.

Such *composite* gels have been useful for the electrophoretic separation of nucleic acids (both RNA and DNA), and for nucleoproteins and other very large proteins that are too large to penetrate the pores of polyacrylamide gels.

To summarise what we have been discussing in this section, you should try answering this SAQ:

SAQ 3.3c

Indicate whether each of the following statements is TRUE or FALSE:

	TRUE	FALSE
(*i*) In polyacrylamide gels, the higher the concentration of polymer the smaller the average pore size of the gel		
(*ii*) A single band in polyacrylamide disc gel electrophoresis proves that the sample consists of only one component		
(*iii*) For the accurate determination of the relative molecular mass of a polypeptide, the method of choice is disc gel electrophoresis		
(*iv*) Agarose gel electrophoresis is useful for the separation of molecules that are too large to enter the pores of polyacrylamide gels		

⟶

SAQ 3.3c (cont.)

	TRUE	FALSE
(*v*) The relative molecular masses of RNA molecules can be determined using polyacrylamide gel electrophoresis		

Summary

The characteristics of the principal types of support media used in zone electrophoresis have been described. Their advantages and disadvantages were discussed, and were related to their use for particular types of molecular separation.

Objectives

When you have completed this Part you should be able to:

- list the types of support media most commonly used in zone electrophoresis;
- describe the main features of these different support media;
- describe, with the aid of diagrams, the electrophoretic apparatus used with different types of support medium;
- outline the applications of the different types of electrophoretic support media;
- discuss the advantages and disadvantages, and the experimental limitations, of different types of support medium used in electrophoresis.

4. Factors that Affect Electrophoretic Mobility

The electrophoretic mobilities of charged molecules are affected by a number of different factors, some of which have been mentioned in earlier Sections of this unit. We will now discuss these factors in more detail.

4.1 CHARACTERISTICS OF THE CHARGED MOLECULES

The electrophoretic mobility of a molecule depends both on its net charge and on its size, which includes its shape as well as its relative molecular mass. Molecules with a high net charge will tend to move more quickly than those with a low net charge under the influence of an applied electric field. However, for molecules with the same net charge, small ones will usually move more quickly than larger ones, because the frictional forces opposing movement are greater for larger molecules. When we are considering the optimum conditions for an electrophoretic separation, we must take these factors into account.

The parameter that combines these two factors is the *charge-to-size ratio* of the molecule. The higher its charge-to-size ratio, the faster should a molecule migrate under given electrophoretic conditions. In Section 1.1. we stated that the migration velocity (v) of a molecule is proportional to its net charge, Q, and inversely proportional to its radius, r. Therefore we can write:

$$v \propto Q/r$$

The term Q/r is the charge-to-size ratio of the molecule.

For mixtures of polynucleotides (see Section 1.2.2) and for proteins treated with sodium dodecyl sulphate (see Section 3.3.2(*b*)) all the molecules have the same charge-to-size ratio, and the electrophoretic separation of components of such mixtures is possible only if the electrophoresis is carried out using sieving gels such as polyacrylamide or agarose. If the migration rate of a molecule is compared with those of molecules of known relative molecular mass under the same electrophoretic conditions, its relative molecular mass can be calculated.

4.2. CHARACTERISTICS OF THE ELECTROPHORETIC SYSTEM

Having considered the features of the substances to be separated, there are a number of parameters of the electrophoretic system itself that we must consider, whose variation can lead to differences in resolution.

The important ones are:

(*i*) for zone electrophoresis, the type of support medium chosen, and, if it is a gel, its pore size (ie gel concentration)

(*ii*) the pH of the electrophoresis buffer

(*iii*) the ionic composition of the electrophoresis buffer

(*iv*) the applied voltage

(*v*) the temperature.

Let us now discuss each of these.

4.2.1. The Support Medium

In Part 3 we have discussed the different types of support medium used in zone electrophoresis. Each of them is suitable for certain applications, and for some analyses there may be several that could be used. Selection of the most suitable one is based on the following considerations:

(*i*) the *size* of the molecules to be analysed, for example whether they are small, such as nucleotides, amino acids and peptides, or large, such as double-stranded DNA and proteins. If a sieving gel is to be used, its pore size (ie its concentration) can be tailored to suit the size of the molecules under investigation.

(*ii*) the *quantity* of sample available for assay

(*iii*) the *cost* of the support medium and the equipment to be used

(*iv*) the *availability* of suitable equipment

(*v*) the *purpose* of the analysis, eg is it to determine whether or not a sample is homogeneous, or is it to determine the relative molecular masses of the components of a sample?

(*vi*) the *time* it would take to run the analysis

(*viii*) the *expertise* of the operator

To give yourself some practice in this type of decision making, try this SAQ:

SAQ 4.2a

For each of the purposes (*i*)–(*v*), decide on an appropriate electrophoretic support medium, and enter your choice on the lines provided:

	Purpose	Support Medium
(*i*)	analysis of a mixture of DNA molecules of M_r 50 to 100 × 10^6	______________
(*ii*)	analysis of a sample of glutamic acid to check for its contamination with glutamine	______________
(*iii*)	accurate determination of the relative molecular masses of the two polypeptide chains of human insulin	______________
(*iv*)	separation of the isoenzymes of lactic dehydrogenase (eg from the serum of a patient with a myocardial infraction)	______________
(*v*)	determination of the number of different coat proteins of a virus	______________

4.2.2. The pH of the Electrophoresis Buffer

For nucleotides and polynucleotides in solution, the net charge is always negative because of the very acidic phosphate groups forming the backbone of these molecules (see Section 1.2.2). The higher the pH the greater the charge on these molecules, and the faster will they migrate in an electric field. Since the charge-to-size ratio of all polynucleotides is virtually the same, they are separated on the basis of their sizes, and changing the pH of the electrophoretic buffer does not affect their separation.

In contrast, you have already seen that pH can markedly affect the net charge on a protein molecule (SAQ 1.2b). As we discussed earlier, (Section 1.2.1), for amino acids, polypeptides, and proteins, there is a pH value, known as the isoelectric point (pI), where the net charge on the molecule is zero. At pH values lower than their isoelectric points, these molecules will have a net positive charge, whereas at pH values above their isoelectric points they will have a net negative charge. At any particular pH value, different proteins in a mixture are likely to have different net charges. If the proteins in a sample are of similar shape and size, we should be able to choose a buffer pH that will give optimum separation of them on the basis of their net charges.

Sometimes two or more proteins can migrate together to give only one band. The appearance of extra protein bands at different buffer pH values signifies the presence of more proteins in the sample than would have been deduced had the analysis been conducted at one pH value only.

An important point to remember is that the pH of buffer solutions depends upon their *temperature*, which is one reason why it is important to minimize variations in temperature during an electrophoretic run. Otherwise, quite significant changes can take place in the pH of the electrophoresis buffer of which the analyst may be unaware, and since these take place during the run they can lead to significant anomalies in the separations achieved.

Furthermore, since the pH of some buffer systems depends on their *concentration*, it is important always to check the pH before use, especially if it is being diluted from a stock solution of more concentrated buffer.

4.2.3. The Ionic Composition of the Electrophoresis Buffer

Interactions can take place between ionisable groups on the surface of charged molecules and ions in the buffer used, such as Na^+, Ca^{2+}, Mg^{2+}, Cl^-, and PO_4^{3-}. The result is that a charged macromolecule becomes surrounded by an ionic atmosphere of opposite charge, so that both its net charge and its electrophoretic mobility are decreased.

Both polypeptide and polynucleotide molecules are susceptible to interactions with ions in buffer solutions. The effect is particularly important in the electrophoretic separation of protein molecules, since different proteins have different amino acid side chains which interact to varying degrees with the ions in the solutions used.

In general, it is advisable to keep the ionic strength of the electrophoresis buffer as low as possible to minimise these 'counterion' effects. However, electrophoresis of many polypeptides and polynucleotides must be carried out in solutions of high ionic strength, otherwise these macromolecules will not be soluble. While increasing the salt concentration increases their solubility, it also changes their net charges and decreases their electrophoretic mobility. Obviously a compromise is necessary in choosing a suitable salt concentration, and usually trial and error is the only way to reach a decision. It is important always to record the salt concentration used for an electrophoretic analysis.

Some proteins may be insoluble at very low salt concentrations, then dissolve as the salt concentration is increased, and finally precipitate out as the salt concentration is increased further. This effect, known as the *salting out* of proteins, varies markedly from protein to protein, and also depends on the nature of the ions (both anions and cations) in the buffer. It can be important if the protein concentration becomes too high, as can happen in disc electrophoresis; it can also occur if there is a large amount of evaporation of water from the buffer during the electrophoresis, leading to an increase in ionic strength.

4.2.4. The Applied Voltage

The source of electrical power used in electrophoresis is usually designed to deliver either constant voltage or constant current. In the literature, the electrical conditions for a particular analysis may be given in terms of current (eg in disc electrophoresis) or applied voltage (eg in paper or agarose gel electrophoresis), where it is often given as the field strength, E V cm^{-1}.

When the power is given in terms of applied voltage, this is not the voltage across the support medium. An electrophoresis system can be considered to be several different resistances in series, the electrical current flowing from one to the next through the system. The resistances are:

- the leads from the power supply
- the reservoirs of buffer solution
- the wicks, if used
- the support medium.

The overall resistance of the system is the sum of all these individual resistances, and consequently the resistance of the support medium is less than the total resistance of the system.

We know from Ohm's law that

$$V = i \times R,$$

where V is the potential difference, or voltage drop, between the ends of a resistance (R), and i is the current flowing through it.

If the current in a system is kept constant, the voltage drop across any component will be proportional to the resistance of that component. Consequently, the voltage drop across the support medium is less than that across the whole system.

Since it is difficult to know exactly what the voltage difference is at the support medium, and since it is likely to vary between runs, it is usually better to operate at constant current.

The migration velocity of a molecule is proportional to the field strength (E V cm^{-1}) across the medium,

$$\text{ie } v \propto E \tag{1.1a}$$

The higher the applied voltage (V), the larger the field strength (E) across the medium, and the faster will a molecule migrate. So increasing the applied voltage will cause the charged molecules to migrate more quickly. This can be an advantage, as it saves time and reduces diffusion of the migrating molecules.

However, as the voltage increases so does the current. An amount of power is generated that is proportional to the current:

$$\text{Power} = i^2 \times R \text{ watts}$$

where i is the current and R the resistance

Some of this power is dissipated as heat (known as *Joule's heat*). A relatively small increase in current results in a large increase in power, and the heat produced can become quite significant. This can have serious effects on an electrophoretic separation, as discussed in Section 4.2.5.

So with applied voltage, just as with the ionic composition of the electrophoresis buffer, it is necessary to compromise. The applied voltage (or current) must be large enough to allow rapid migration of charged molecules, but not so large that excessive heat is generated.

4.2.5. Temperature

If Joule's heat generated during an electrophoretic run is not removed by an efficient cooling system, there are a number of possible effects that it can have. We will consider these individually, but it must be remembered that several of them may occur simultaneously, which will obviously complicate interpretation of the results.

(a) *Convection Currents*

When Tiselius was developing the technique of moving boundary electrophoresis, he noticed that convection currents appeared in the U-tube apparatus (see Section 2.1), which adversely affected the analyses. Warmer solution in the centre of the apparatus had a lower density than cooler solution close to the walls, and this difference caused the convection currents. Since the maximum density of water is at 4 °C, and around this temperature the density of aqueous solutions shows least variation with change in temperature, it is advisable to run electrophoretic analyses as close as possible to 4 °C to avoid the density differences that lead to convection currents. To achieve this, an efficient method of cooling should be incorporated, or if this is not possible either the system should be placed in a refrigerated room (a 'cold room'), or the buffer solution should be kept cold by immersing the reservoirs in ice-baths.

(b) *Diffusion*

The diffusion of migrating zones of charged molecules increases with an increase in temperature. If the electrophoresis takes several hours to run, diffusion effects can become very significant.

(c) *Distortion of Zones*

Temperature is a very critical factor in polyacrylamide gel electrophoresis, particularly in column gels. If cooling is inadequate, those parts of the migrating zones in the warmer part of the gel (ie the centre) will move faster than those in the cooler part (ie the outside). This unequal speed of migration results in curved bands, which can lead to overlap between neighbouring zones, and consequently very poor resolution.

(d) *Evaporation*

Electrophoresis is always carried out in a closed system, to minimise loss of water by evaporation, which increases with temperature. This can lead to drying out of some types of support medium (eg cellulose acetate), and it can lead to an increase in the ionic strength of the buffer during the analysis (see Section 4.2.3).

(*e*) *Viscosity*

An increase in temperature during gel electrophoresis can change the viscosity of the gel, making it softer. This is particularly important with agarose gels. As this happens during the electrophoretic run, its effects on the results obtained are complex, hence it is important to keep the temperature of the system constant by efficient cooling.

We have already decided that 4 °C is a good temperature for an analysis. However, the viscosity of aqueous solutions increases as their temperature decreases. This means that the frictional resistance to migration of charged molecules will increase, and if the temperature of the electrophoresis system is maintained at 4 °C, the electrophoretic mobilities of charged molecules will be relatively low.

Therefore, we must compromise, and choose a temperature that is optimal for the particular analysis we want to carry out. The most important point is that we should make every effort to keep the temperature *constant*, whatever temperature we decide to use.

To summarise what we have discussed in this Section, and to see if you can apply what you have learnt, try answering these two SAQs:

SAQ 4.2b

In determining the conditions for the maximum electrophoretic separation of two proteins, which parameter is likely to have the *greatest* effect?

Choose one of the following:

(*i*) pH
(*ii*) ionic strength
(*iii*) temperature
(*iv*) current

SAQ 4.2c

In the electrophoretic separation of several molecules of double-stranded DNA, which parameter is likely to have the *greatest* effect?

Choose one of the following:

(*i*) pH
(*ii*) buffer ionic composition
(*iii*) temperature
(*iv*) gel pore size

Summary

The different factors that affect electrophoretic separations have been discussed in this Section. The characteristics of the charged molecules in the sample, and various features of the electrophoretic system, have all been considered.

Objectives

When you have completed this Part you should be able to:

- outline the characteristics of charged molecules that affect their separation by electrophoresis;
- identify in an electrophoretic system those features that are likely to influence the separation of charged molecules;
- discuss how various factors affect the separation of charged molecules by electrophoresis.

5. Detection of Sample Components Separated by Electrophoresis

After electrophoresis has been completed we need to locate the separated components of the sample. Exactly how we would do this depends on:

- the nature of the molecules to be detected
- the type of electrophoresis system used
- the purpose of the analysis.

We will now discuss the main kinds of detection methods that have been used in electrophoresis.

5.1. OPTICAL METHODS

When electrophoresis was originally developed, using the moving boundary method, the separated proteins were detected by photographing the electrophoretogram under ultraviolet (UV) radiation. The positions of the protein boundaries showed up as dark bands on the photographic plate. In later work, the detection method used was the Schlieren optical method, which detects local changes in refractive index and allows visualisation of the separating components while the separation is actually occurring.

Both of these methods depend upon the interaction of molecules with electromagnetic radiation, and this has been the basis of a number of detection techniques used in zone electrophoresis. Many of these involve staining techniques, which use either absorption or fluorescence methods to detect the stained molecules.

5.1.1. UV Absorption by Separated Components

Most proteins absorb UV radiation, and have absorbance maxima between 230 nm and 280 nm. The molar absorptivity of a protein molecule depends on the proportions of the amino acids tryptophan, tyrosine and phenylalanine present, all of which absorb UV radiation. The absorption spectra of these three amino acids are shown in Fig. 5.1a:

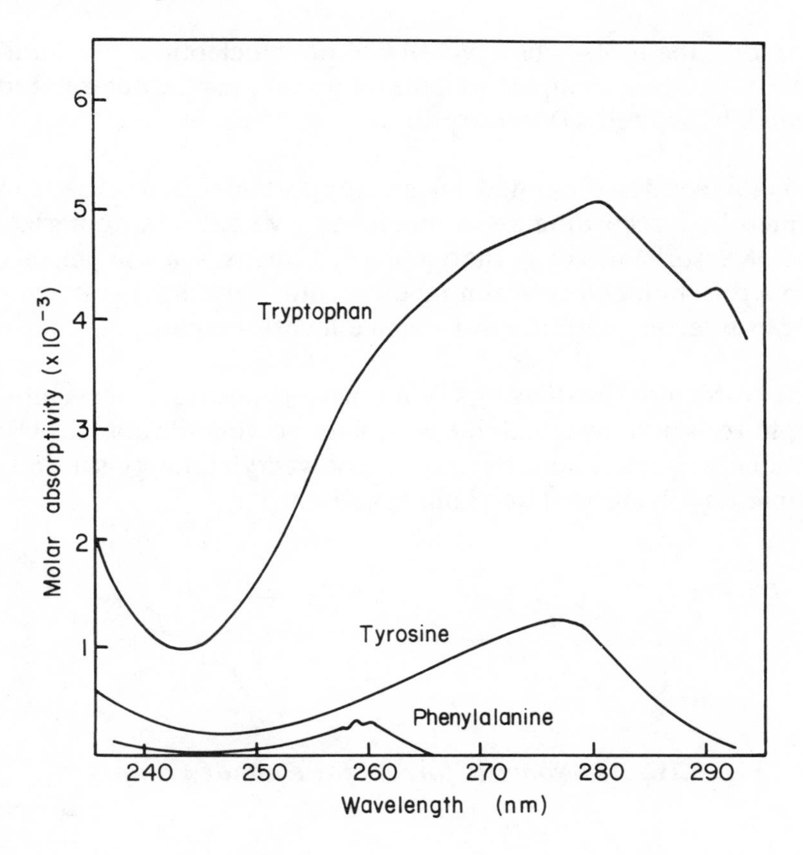

Fig. 5.1a. *Absorption spectra of tryptophan, tyrosine and phenylalanine*

Although proteins have a much higher absorbance at 230 nm than at 260 nm or 280 nm, due to resonance of the peptide bonds, it is usually not practicable to use this wavelength because of interference from other materials in the system, including the support medium itself.

Polynucleotides, including RNA and DNA, absorb UV radiation strongly, with a maximum absorbance at 260 nm. This is due to resonance in the heterocyclic ring (see SAQ 1c, *Study Guide*). Since the molar absorptivities of polynucleotides at 260 nm are much higher than those for proteins at 280 nm, the sensitivity of using UV absorption as a method of detection is much greater for nucleic acids than for proteins.

Furthermore, the molar absorptivities of polynucleotides are independent of their base composition, therefore they can be determined quantitatively by their UV absorption.

For polynucleotides separated on gel support media, the gel can be scanned by a recording densitometer to give either a numerical readout of absorbance or a chart record. Usually agarose gels are dried to a thin film before scanning them, otherwise light scattering effects can interfere with the absorbance measurements.

There are difficulties with using UV absorbance measurements after electrophoresis in polyacrylamide gels, since acrylamide absorbs UV radiation because of its amide group and polyacrylamide gels usually have some unpolymerised acrylamide present.

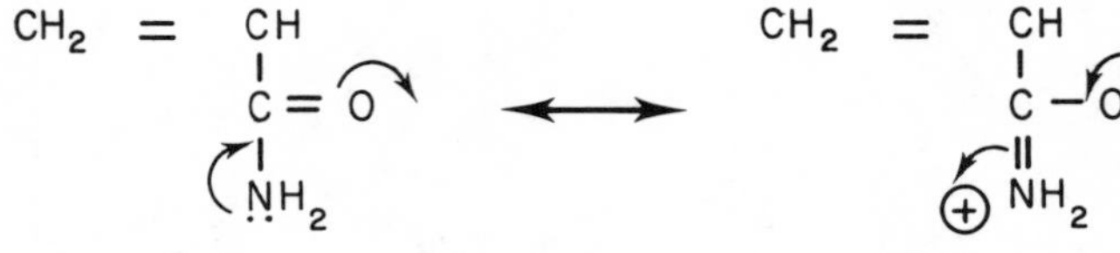

$CH_2=CH-C(=O)-\ddot{N}H_2 \longleftrightarrow CH_2=CH-C(-O^{\ominus})=N^{\oplus}H_2$

Fig. 5.1b. *Mesomeric forms of the amide group*

Therefore the 'background' absorbance of these gels at wavelengths below 270 nm can be quite high. This can make the detection of proteins difficult, but polynucleotides can usually be detected because of their high molar absorptivities.

5.1.2. Staining

Probably the most widely used method of detection is to react the separated components with a stain or dye then detect their positions by absorption or fluorescence. Staining methods have been used with support media of all kinds.

Π Can you think of some features that would be desirable in chemicals used for staining molecules after an electrophoretic separation?

There are several features that you might have chosen, among them being:

(*a*) the chemical should be *selective* for the type of substance under investigation,

(*b*) the product formed between the chemical and the substance of interest must be *stable* and *insoluble* during the staining and destaining procedures,

(*c*) the chemical should *react quickly* with the substances of interest and it must be possible to remove any unbound chemical quickly and completely,

(*d*) the stained product should have a *high molar absorptivity* or *fluorescence emission*, to ensure high sensitivity.

We can choose to stain substances either:

(*i*) *before electrophoresis*, by reacting them with a small amount of the staining reagent. In this case, the migration of the stained molecules can be followed during the run.

(*ii*) *after electrophoresis*, by spraying the support medium (eg paper) or by immersing it in a staining solution. When the applied voltage is switched off at the end of the run, diffusion of the separated molecules continues, and this leads to loss of resolution. To minimise this, it is usual to *fix* the separated components *in situ* immediately after stopping the run. This is done either by drying the support medium (if it is paper or cellulose acetate), or by precipitating the molecules within the support medium (if it is a gel), using a chemical *fixative* such as trichloroethanoic acid or ethanoic acid.

After an electrophoretogram has been stained, excess unbound stain must be removed before the stained molecules can be detected. We can achieve this either by extensively washing or soaking the support medium in *destaining* solution. This is usually a slow procedure, though it can be speeded up by stirring, and by changing the destaining solution frequently.

A more efficient, and faster, destaining method is *electrophoretic destaining*, in which the support medium (usually a gel) is put into a specially designed apparatus which allows electrophoresis to be carried out at right angles to the direction of that of the original separation. Under the influence of an applied voltage, unbound stain moves out of the gel leaving fixed and stained molecules within it. A typical system is shown in Fig. 5.1c:

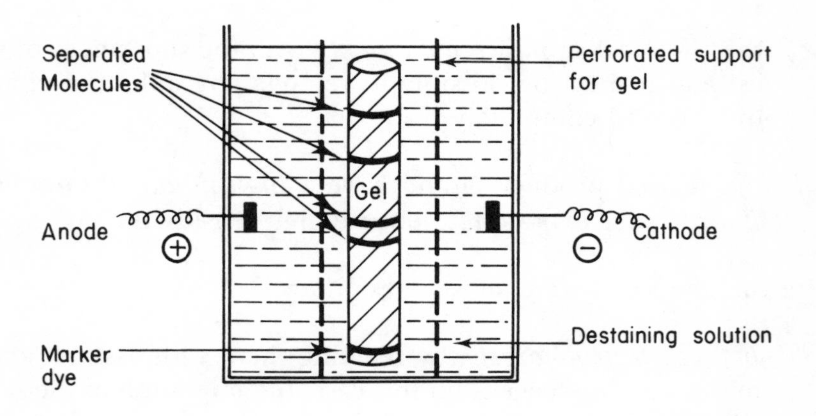

Fig. 5.1c. *Apparatus for electrophoretic destaining*

Many different chemicals have been used to stain the separated components.

Specific stains may be used to detect substances bound to proteins. For example, some metalloproteins can be localised by a colour reaction specific for the metal, eg staining of copper-containing proteins (Fig. 5.1d).

Types of substance stained	Staining reagent	Comments
amino acids, peptides, and proteins	Ninhydrin	Very sensitive stain for amino acids, either free or combined in polypeptides. Used after paper electrophoresis.
proteins	Amido Black 10B	Binds to cationic groups on proteins. Adsorbs onto cellulose, giving high background staining with paper and cellulose acetate. Destaining causes dehydration and shrinkage of polyacrylamide gels.
	Coomassie Brilliant Blue	Binds to basic groups on proteins, and also by non-polar interactions. Widely used stain.
	Ponceau S (Ponceau Red)	Used routinely in clinical laboratories for cellulose acetate and starch gels. Very rapid staining reaction which leaves a clear background.
glycoproteins	Alcian Blue	Stains the sugar moiety.
copper-containing proteins	Alizarin Blue S	Specifically indicates the presence of copper.
polynucleotides, including RNA and DNA	Acridine orange	Stained product can be assessed quantitatively.
	Pyronine Y (or G)	Gives a permanent staining, so electrophoretogram can be stored for several weeks.
proteins, lipids carbohydrates, polynucleotides	Stains-All	Wide applicability, as it forms characteristic coloured products with many different types of molecule. Low sensitivity

Fig. 5.1d. *Chromogenic stains*

As well as being used to detect the separated components, some stains can be used quantitatively, the intensity of the stained band being proportional to the amount of the separated material present, eg acridine orange for staining polynucleotides.

There are many different stains available that give either coloured or fluorescent products, ie chromogenic and fluorescent stains. In choosing an appropriate stain, the nature of the molecules to be detected is a major consideration.

Some of the chromogenic stains most widely used are listed in Fig. 5.1d.

When using fluorescent staining techniques, the sample is usually treated with the dye *before* electrophoresis, and the migrating substances are detected by viewing the electrophoretogram in radiation of the appropriate wavelength (usually under a UV lamp). After electrophoresis, the support medium can either be photographed in a suitable light, or scanned by a recording fluorimeter. Only very small amounts of dye should be used, to ensure that the electrophoretic mobilities of the stained molecules are not altered by the presence of the stain.

The fluorescent stains most widely used are listed in Fig. 5.1e:

Types of substance stained	Staining reagent	Comments
proteins	Dansyl chloride	Reacts with amine groups
	1-Anilino-8-naphthalene sulphonic acid (ANS)	Non-fluorescent, but gives fluorescent product
	Fluorescamine	Non-fluorescent, but gives a fluorescent product
polynucleotides, including RNA and DNA	Acridine orange	
double-stranded polynucleotides	Ethidium bromide	Very sensitive. Widely used with agarose gels

Fig. 5.1e. *Fluorescent stains*

While this is not a comprehensive list of the many stains available, Figs. 5.1d and 5.1e describe the stains that you are most likely to find recommended. To revise the main points, you should try this SAQ:

SAQ 5.1a

Select from the stains listed below the one that is most appropriate for each of the purposes (*i*)–(*vi*), and enter your choice in the spaces provided:

(*i*) staining of proteins in polyacrylamide gels ______________

(*ii*) detection of small peptides after paper electrophoresis ______________

(*iii*) fluorescent labelling of double stranded DNA molecules ______________

(*iv*) identification of bands of glycoproteins after electrophoresis of a mixture of proteins ______________

(*v*) rapid detection of proteins separated on cellulose acetate membranes ______________

(*vi*) fluorescent tagging of proteins in SDS-polyacrylamide gels ______________

Alcian Blue
Amido Black
Coomassie Brilliant Blue
Ethidium Bromide
Fluorescamine
Ninhydrin
Ponceau S
Pyronine Y

5.2. RADIOCHEMICAL METHODS

Radiochemical methods of detection are widely used as an alternative, or in addition, to optical methods. Before electrophoresis, substances are labelled with a radioactive isotope. After electrophoresis, the support medium may be cut into pieces and the radioactivity in each piece measured by liquid scintillation counting methods. Alternatively, the technique of autoradiography may be used, which allows detection of the radioactivity in the separated components without the need for cutting up the support medium.

Considerable thought must be given to deciding which radioisotope to use. High-energy emitters are extremely useful labels to use for autoradiography, especially with gel electrophoresis. Proteins may be labelled with the readily-detected γ-emitter ^{125}I. The high-energy β-emitter, ^{32}P, has been widely used to label nucleic acids. ^{14}C and ^{35}S are medium-energy β-emitters and ^{3}H is a low energy β-emitter. These isotopes have all been used widely as tracers in electrophoretic separations. The point to remember is that the higher the energy the more easily is the radioactive decay detected, both by liquid scintillation counting and by autoradiography. If you want short counting times (in liquid scintillation counting) or short exposure times (in autoradiography) then you should choose the highest energy radioisotope that provides an appropriate label for the substances under investigation.

SAQ 5.2a

From the radioisotopes given below, choose the one you think would be the *most* appropriate for rapid assay of the various samples after gel electrophoresis. Put your choice on the lines provided. ⟶

SAQ 5.2a (cont.)

(*i*) Proteins containing several methionine residues ____________

(*ii*) Newly-synthesised RNA molecules ____________

(*iii*) Proteins containing the amino acid tyrosine ____________

(*iv*) New proteins synthesised in cells following viral infection ____________

^{35}S ^{32}P ^{125}I ^{3}H ^{14}C

5.2.1. Liquid Scintillation Counting

If the support medium is paper or cellulose acetate, it is dried after electrophoresis, cut into strips of the selected size, and the pieces put directly into a suitable liquid scintillation counting mixture. Polyacrylamide and agarose slab gels can also be dried, using vacuum-suction (see Section 3.3.3), and these can then be cut into sections. However, samples in gel support media usually need to be solubilised before their radioactivity can be measured by liquid scintillation techniques. There are several methods used for solubilising the samples:

(*i*) they may be eluted from the slices by an appropriate buffer solution, or an organic reagent eg hyamine hydroxide, piperidine, or a commercially available *solubiliser*.

(*ii*) the gel structure may be destroyed using sonication, or a chemical agent such as formamide (for agarose gels) or 30% aqueous hydrogen peroxide (for polyacrylamide gels).

Care must always be taken to choose a liquid scintillation counting mixture that is compatible with the solubilisation method used.

Starch and some agarose can be sliced into zones using a sharp razor blade, but polyacrylamide gels require special devices to cut them into small and even slices without distorting the zones. One of the simplest and most commonly-used types of apparatus consists of a set of sharp blades evenly-spaced along a support, as shown in Fig. 5.2a:

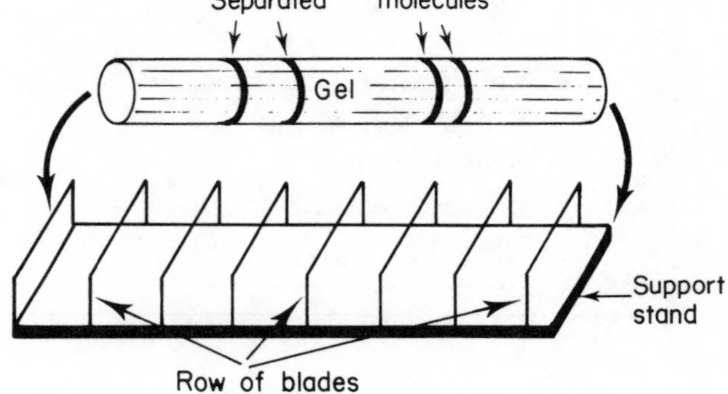

Fig. 5.2a. *Apparatus for slicing polyacrylamide tube gels*

Guillotine-type gel slicers are also commonly used, especially after freezing the gel in dry ice to minimise distortion of the zones during cutting.

Polyacrylamide gels may also be sliced along their length to give two or more slices that can be examined by different detection methods.

5.2.2. Autoradiography

The technique of autoradiography is useful for detecting radioactivity in labelled compounds separated by zone electrophoresis. It has been exceptionally useful for gels of large pore size, such as dilute agarose gels,since these are very soft and difficult to handle for liquid scintillation counting.

The first step is to dry the support medium – if it is a slab gel, vacuum suction is generally used until the gel forms a transparent film. The dried support medium is then put into a specially designed photographic cassette (Fig. 5.2b), available from several commercial suppliers, where it is held firmly in place by pressure against a sheet of photographic film (usually X-ray film). A thin protecting sheet may be placed between the sample and the photographic emulsion to prevent contact of the film with chemicals from the sample which might increase the 'background' of the autoradiograph. Any radiation which passes through the photographic film is reflected as fluorescent emission by an intensifying screen, and this enhances the photographic image.

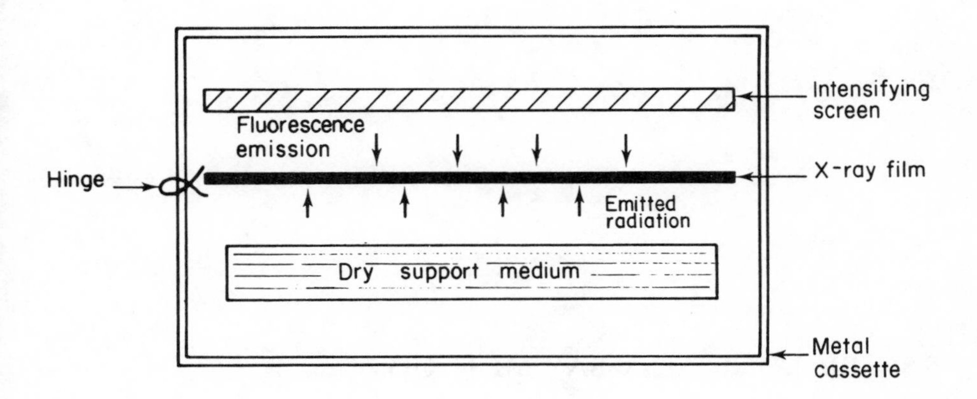

Fig. 5.2b. *Autoradiography cassette*

The loaded cassette is left at a temperature below O °C for an appropriate time (usually several days), and when the film is developed the distribution of radioactivity in the sample shows up as black areas on a light background.

SAQ 5.2b

Suggest *three* factors which would affect the sensitivity of detection of radioactivity in labelled molecules after electrophoresis.

5.3. BIOLOGICAL ASSAY METHODS

For the location of proteins after electrophoresis, biological assay methods have been of great value.

Π Can you suggest what might be the main advantage of biological assay methods for proteins, compared to optical or radiochemical detection methods?

If you suggested that biological assay methods are likely to be very *selective* for individual proteins, well done! This is probably their main advantage. Another correct answer you may have given is that biological assay methods are often very *sensitive*, and can detect much smaller amounts of the particular protein of interest than can most other methods of detection.

There are two types of biological activity of proteins which have been especially useful for their detection after electrophoresis. These are:

- their enzymic activity (if they are enzymes), and
- their antigenic (immunological) characteristics (dealt with in Part 6).

The most generally applicable method for the detection of enzymes involves cutting the support medium into small pieces, eluting the enzyme(s) from the individual fragments, and then determining enzyme activity. However, from the point of view of convenience, and also for improved resolution, it is preferable to assay enzyme activities *in situ*, ie without cutting up the support medium into fragments. Gels sliced longitudinally have often been used in this way.

The methods used for detecting the activities of enzymes after electrophoresis are based on:

- methods used originally for the histochemical localisation of enzymes in cells and tissues, and

- methods developed originally for the assay of enzyme activity in solutions.

What features of an enzyme assay system are likely to be important if it is to be used to detect enzymes after electrophoresis? Firstly, the reaction that is used as the basis of the enzyme assay system must be catalysed specifically by the enzyme of interest. Also, care must be taken to avoid loss of enzyme activity of the protein, both during electrophoresis and during the enzyme assay. Buffered solutions of appropriate pH must be used, and care must be taken to avoid temperatures that could be harmful to heat-sensitive enzymes. Chemicals present in the electrophoresis system may be inhibitory to enzyme activity, particularly in some gel electrophoresis systems, and we should remember that anything that denatures a protein, for example SDS, is going to destroy enzyme activity.

If the assay is to be used for the detection of enzyme activity *in situ*, we must choose one that will produce a *visible change* at the location of the enzyme. There are several ways of doing this:

(*i*) We can choose an assay reaction for which the product has a distinctive *colour*. The appearance of a coloured band on the electrophoretic support medium indicates the location of the enzyme assayed.

An example of this is the use of nitrophenyl derivatives as substrates for various hydrolases – the appearance of the bright yellow product, 4-nitrophenol, localises these enzymes. For phosphatases, 4-nitrophenyl phosphate is commonly used as a substrate:

$$O_2N-C_6H_4-PO_4 \xrightarrow[\text{Phosphatase}]{H_2O} O_2N-C_6H_4-OH + H_3PO_4$$

4-Nitrophenyl phosphate → 4-Nitrophenol (yellow)

(*ii*) For some enzymes, the product of the enzyme-catalysed reaction may be *fluorescent*.

(*iii*) Many enzymes are located after electrophoresis in gels by *coupling* the colourless product of the enzyme-catalysed reaction with a suitable stain, so that a coloured dye is formed. An example of this is the detection of esterases:

$$\text{1-naphthylacetate} \xrightarrow[\text{esterase}]{H_2O} \text{1-naphthol} + \text{ethanoic acid}$$

$$\text{1-naphthol} \xrightarrow{\text{Fast Red}} \text{Coloured azo dye}$$

(*iv*) In some cases the colourless product of an enzyme-catalysed reaction is converted by means of an *auxiliary enzyme* to yield a coloured product.

A very large number of enzymes have been detected by their biological activities *in situ* after electrophoresis on different support media. Detection of enzyme activity has been particularly important for isoenzymes, which have different migration rates but which have the same biological activity.

Summary

In this Part, you have been introduced to the techniques most commonly used for the detection of substances after their separation by electrophoresis. Three main classes of detection method have been described – optical methods, radiochemical methods, and biological assay methods.

Objectives

You should now be able to:

- outline the principles of the different detection methods used in electrophoresis;
- decide which type of detection method would be most suitable for the detection and identification of substances after their separation by an appropriate electrophoretic system;
- discuss the advantages and disadvantages of the different types of detection method used in electrophoresis;
- select from a list a stain appropriate for use in a given electrophoretic separation;
- devise a biological assay method for use in the detection of an enzyme, of known reactivity, that has been subjected to electrophoresis;
- interpret results from electrophoretic separations where separated substances have been detected by various methods.

6. Immunoelectrophoresis

We mentioned in Section 5.3 that biological assay methods are very selective and sensitive for detecting proteins separated by electrophoresis. If the protein of interest is not an enzyme, or if there is no suitable assay for its enzymic activity, or if it is present in an amount too small to be detectable using an enzyme activity assay method, immunodetection methods may be suitable. These methods are based on the ability of the protein of interest to elicit the production of an antibody in an animal (ie to behave as an *antigen*). Provided that the antibody can be isolated, it will react specifically with the protein that elicited it, and not with any other.

A variety of techniques that combine electrophoresis and immunology has been devised, all of which depend on the formation of a visible precipitate between an antigen and its specific antibody. These techniques have been useful for the analysis of small amounts of individual proteins in mixtures, and for checking the purity of protein preparations by confirming the presence of only one antigen.

For each antigen–antibody system, there is an optimum ratio of antibody to antigen at which the antibody–antigen complex will give a visible precipitate, called *precipitin*. If a great excess of either antibody or antigen is present in the reaction mixture, no precipitin will be observed. This phenomenon is illustrated in Fig. 6.a, which is called a *precipitin curve*. The region where maximum precipitin is formed is known as the *equivalence region* for a particular antibody-antigen system.

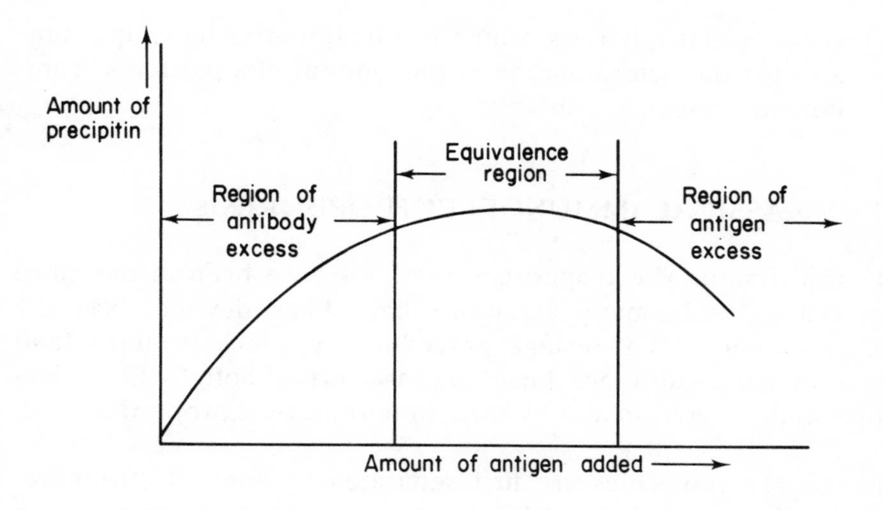

Fig. 6a. *Precipitin curve*

It is important that sufficient time is allowed for all the antibody-binding sites in the antigen to become effective, producing the extensively cross-linked antigen–antibody lattice structure which is seen as precipitin. This time depends on the individual system. Precipitin can usually be detected on an electrophoretogram simply by looking at it under a good light. However, the sensitivity of detection of precipitin can be increased by staining it with a protein stain, such as Ponceau S, or by using a radioactively labelled antigen or antibody.

Another immunochemical 'trick' is to add to the antibody–antigen reaction mixture a *second antibody* directed against the first antibody. This second antibody may be radioactively labelled (say with ^{125}I) to further increase the sensitivity of detection.

There are two main types of immunoelectrophoresis:

(*i*) *classical* immunoelectrophoresis, or electroimmunoassay, which is useful for the separation and identification of components in mixtures of antigens, such as cell extracts and body fluids. This analytical technique cannot be used for the quantitative determination of antigenic substances.

(*ii*) *rocket* electrophoresis, which is a quantitative technique suitable for the determination of the amount of a particular component present in a mixture.

6.1. CLASSICAL IMMUNOELECTROPHORESIS

Classical immunoelectrophoretic methods have been in use since the 1950s. While many variations have been devised, basically the technique is a two-stage procedure by which proteins (and other antigenic substances) can be characterised both by their electrophoretic migration and by their immunological properties.

The antigen molecules are first separated by zone electrophoresis, usually on a gel of 1–1.5% agar or agarose, but sometimes on other support media, eg cellulose acetate membranes. A typical arrangement for gel immunoelectrophoresis, as seen looking at it from above, is shown in Fig. 6.1a.

The antigen mixture is placed in a hole in the gel and electrophoresis is carried out. The power supply is switched off, and appropriate antiserum is placed in troughs that lie parallel to the direction of the electrophoretic run. The electrophoretogram is placed in a moist chamber at constant temperature. The antigens and the antibodies diffuse into the gel, and each antigen encounters and reacts with its specific antibody in the antiserum. Maximum precipitation occurs where the concentration ratio of antigen to antibody is within its equivalence region – see Fig. 6.1a. Because the antigens are usually smaller than the antibodies, they will diffuse faster through the gel. There will be a leading edge where the antigen will be in excess over antibody, until eventually a concentration of antibody is reached that is optimal for precipitin formation. At this point, the antigen stops diffusing and the precipitin appears as lines, curved towards the antibody troughs.

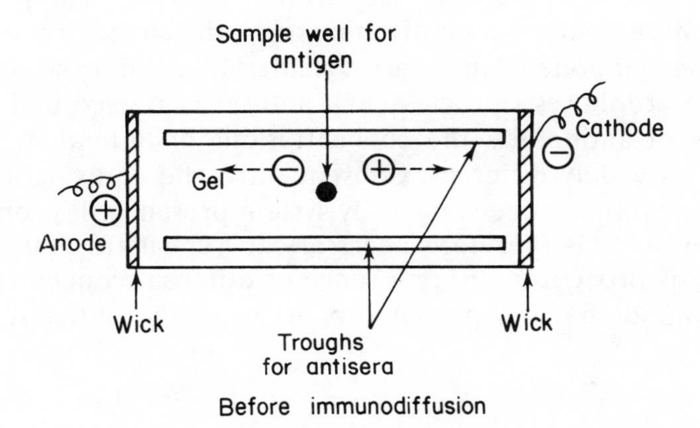

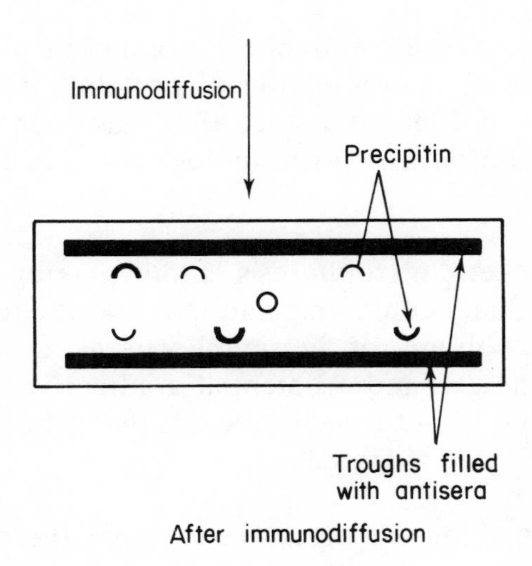

Fig. 6.1a. *Classical immunoelectrophoresis*

When precipitin formation is complete, the gel can be washed with saline solution to remove non-precipitated material. The precipitin lines can then be stained with an appropriate dye.

The information derived from immunoelectrophoresis depends very much on careful choice and use of the antiserum. One problem is that in a multicomponent antigen mixture, the concentrations of the individual components may vary considerably, and these are diluted as the electrophoresis progresses. If antiserum is used that contains a variety of antibodies, the concentrations of individual antibodies may be widely different. Consequently, the equivalence region for each possible antigen–antibody system present will probably not be achieved. It is therefore necessary to repeat the immunoelectrophoresis procedure, using a range of different concentrations of antiserum, so that the presence of none of the antigens is overlooked.

6.2. ROCKET ELECTROPHORESIS

Rocket electrophoresis is an immunoelectrophoretic technique that allows the quantitative determination of proteins. It was first described by Laurell in 1966, and since then has been widely used for the determination of proteins in biological fluids eg serum and cerebrospinal fluid.

The method of rocket electrophoresis entails making a thin sheet of 1–2% agarose gel and containing antibody specific to the protein of interest. Precise volumes of the sample (ie the antigen) are put into wells cut in the gel, and a potential gradient is applied to it. The proteins migrate from the wells towards the cathode, becoming ever more dilute as they migrate.

At the leading edge of the moving antigen sample, there is an excess concentration of antigen relative to antibody, so little precipitin will be formed (see Figure 6.2a). The antigen will migrate through the gel until the concentration of antigen at the leading edge of the migrating sample becomes equivalent to the antibody concentration in the gel. At this point the antigen–antibody complex will form precipitin, and the antigen will migrate no further. Consequently, the higher the concentration of the antigen in the sample well, the further will the sample migrate before its equivalence region is reached. The result is that *rocket-shaped* lines of precipitin are produced, the heights of which are proportional to the concentration of specific antigen originally put into the sample well.

On each plate a series of standard antigen samples of known concentration are applied to 'standard' wells. By measuring the distances between the tips of the 'standard' rockets and the upper edge of their wells, a calibration curve of antigen concentration versus rocket height can be constructed. Using this, the concentration of antigen in the sample wells can be determined.

Fig. 6.2a illustrates the technique. Wells (*a*)–(*e*) contained decreasing amounts of antigen standard solution. Wells (*f*)–(*h*) were loaded with 5 μl samples containing an unknown concentration of antigen. The gel contains a 1% w/v solution of antibody specific to the antigen of interest.

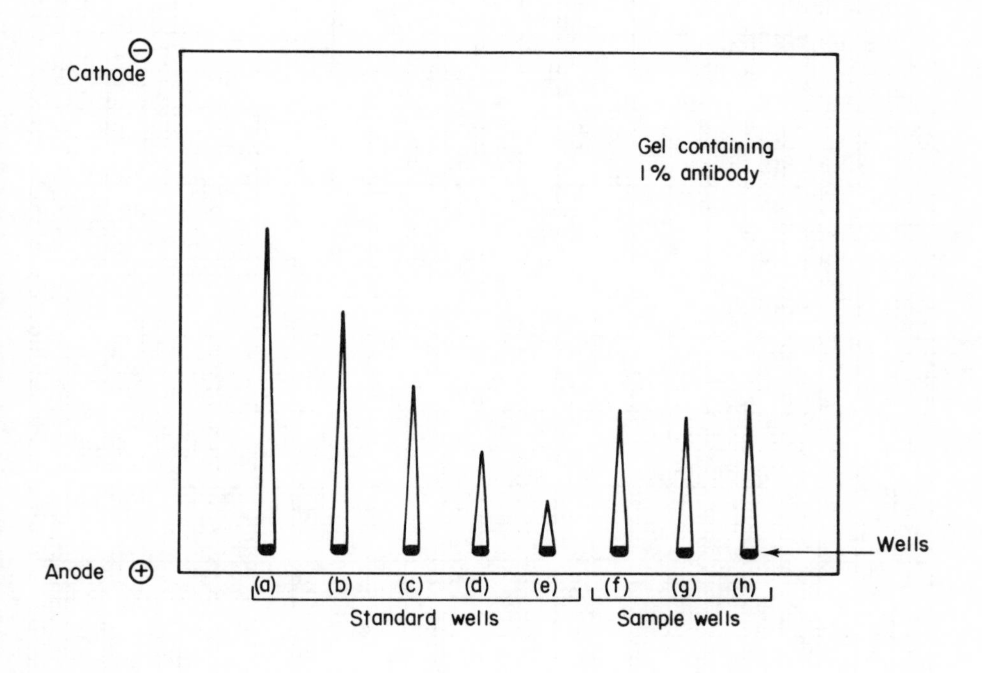

Fig. 6.2a. *Rocket electrophoresis*

SAQ 6.2a

In Fig. 6.2a, the standard reservoirs, (*a*)–(*e*), contained 1.0, 0.75, 0.5, 0.25 and 0.1 μg of antigen respectively. Draw a calibration graph relating rocket height to antigen amount, and from it determine the concentration of this antigen in the sample.

If the sample contains a mixture of substances, only the specific antigen will react with the antibody in the gel to form precipitin. The other antigens can be removed from the gel, after electrophoresis, by washing the gel with saline solution. The lines of precipitin may then be stained with an appropriate dye.

Rocket electrophoresis is an elegant and useful technique. To ensure good results, the electrophoresis tank should be fitted with an efficient cooling system. Also, clearer results are usually obtained if the electrophoresis is carried out with a low field strength (say 2 V cm^{-1}) overnight rather than with a higher one (say 8–10 V cm^{-1}) for only a few hours.

Summary

The fundamental principles of antibody–antigen precipitation were outlined briefly, and the use of immunoprecipitation reactions as the basis of detection methods for molecules separated by electrophoresis were described. Both classical immunoelectrophoresis and rocket immunoelectrophoresis were discussed.

Objectives

After you have studied this Part you should be able to:

- draw and interpret a precipitin curve;
- outline the principles of classical immunoelectrophoresis;
- with the aid of diagrams, describe a classical immunoelectrophoresis system;
- outline the principles of rocket electrophoresis;
- with the aid of diagrams, describe a rocket immunoelectrophoresis system;
- interpret electrophoretograms from classical immunoelectrophoresis and rocket electrophoresis analyses.

7. Two-dimensional Techniques Involving Electrophoresis

For very complex mixtures, separation by electrophoresis in one dimension is often inadequate. Resolution may be substantially improved by combining electrophoresis with a second separative technique. The electrophoresis is carried out in one direction, and the second separative procedure is usually carried out in a direction perpendicular to this. The techniques commonly used in combination with electrophoresis in this way are:

— chromatography,
— electrophoresis under different conditions,
— isoelectric focussing (see Section 2.3.1).

7.1. COMBINED ELECTROPHORESIS AND CHROMATOGRAPHY

High-voltage electrophoresis has been a very useful technique for the separation of amino acids and peptides. However, greatly improved resolution of complex mixtures of these molecules can be obtained by a combination of electrophoresis and chromatography. The sample may first be subjected to a chromatographic separation, then to electrophoresis, or *vice versa*.

7.1.1. Fingerprinting of Proteins

A technique that has been widely used in the characterisation and identification of proteins is *fingerprinting* or *peptide mapping* (Ingram, 1957). To fingerprint a protein, it is partially digested using specific proteolytic enzymes, such as trypsin, which produce small peptides. The peptide digest is then subjected to combined chromatography and electrophoresis, usually carried out on a single support medium (normally paper). A pattern of spots, known as a *fingerprint*, is obtained, each spot being one of the components of the peptide digest. The spots may be detected by examining the fingerprint under a UV lamp, or after staining with ninhydrin (Section 5.1.2). A diagram of a fingerprint is shown in Fig. 7.1a:

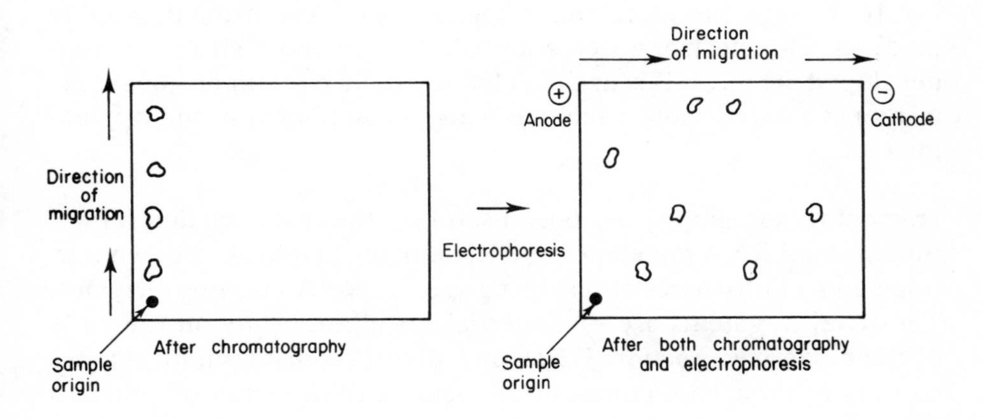

Fig. 7.1a. *Fingerprinting of proteins*

Comparison of the resulting pattern of spots with other 'standard' fingerprints, allows the primary structure (amino acid sequence) of the protein to be deduced.

Fingerprint analysis can reveal subtle differences between the primary structures of closely related protein molecules, eg human haemoglobin variants in which a single amino acid substitution has occured. The part of the protein molecule that differs in amino acid composition can be identified readily from the location of spots on the fingerprints. The nature of the amino acid chain can be identified by eluting the relevant spots from the paper and determining their amino acid compositions.

7.1.2. Homochromatography

In homochromatography, the chromatography is carried out on a support medium different from that used for the electrophoresis which is usually paper or cellulose acetate. The separated components are then transferred to the chromatography support medium, which may be an ion-exchange material such as diethylaminoethyl (DEAE) cellulose. The transfer of material from one support medium to the other is achieved by placing them in firm contact with each other and allowing a buffer or other liquid to pass slowly through the electrophoretogram onto the chromatographic medium. The electrophoretogram is removed and a chromatogram developed. The result is again a characteristic two-dimensional pattern of separated spots, each one being a component of the original mixture.

Homochromatography has been useful for the sequence determination of small RNA molecules, the RNA being first broken down into fragments of fifty bases or less using specific RNA-cleaving enzymes. The RNA fragments are separated electrophoretically on the basis of their size (see Section 1.2.2), and then chromatographically according to their base composition, using DEAE-cellulose either as a paper or a thin-layer. The separated fragments are removed from the DEAE-cellulose and their base compositions determined.

7.2. TWO-DIMENSIONAL ELECTROPHORESIS

In this technique, electrophoresis is carried out twice, the direction of the second being at right angles to that of the first. In the simplest case, the same support medium (usually paper) is used for both operations, and the result is a two-dimensional pattern of spots, each one being a component of the original sample. It is important to change at least one of the experimental conditions for the second run, and this change must affect the individual components in the sample differently, so that their migration rates change relative to one another.

Π Think about this for a minute or two. Do you understand why this is necessary, and what would happen if it were not done?

If you did not change the conditions in this way, the components would move relative to each other in the same way in the second direction as they did in the first. The spots would all end up on a diagonal line (Fig. 7.2a):

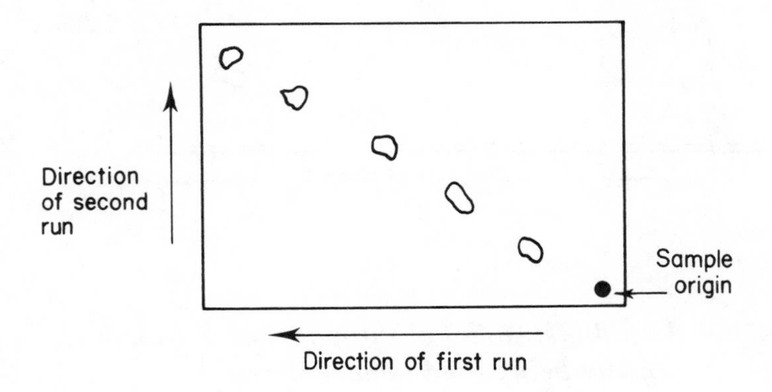

Fig. 7.2a. *Two-dimensional electrophoresis, the same experimental conditions being used in each dimension*

You might as well have done a single run in the direction of the diagonal!

Ideally, the two runs should separate the components according to different parameters eg one run could separate them on the basis of their sizes and the other on the basis of their charges.

If the support medium for the second run is to be different from that used for the first, transfer of samples from one to the other has to be achieved. The initial studies with two-dimensional electrophoresis were carried out using filter paper strips for the first run and starch gels for the second (Smithies and Poulik, 1956). The first electrophoretogram was inserted into a slit cut across the starch gel, and then electrophoresis was carried out in the direction perpendicular to the first run, as shown in Fig. 7.2b:

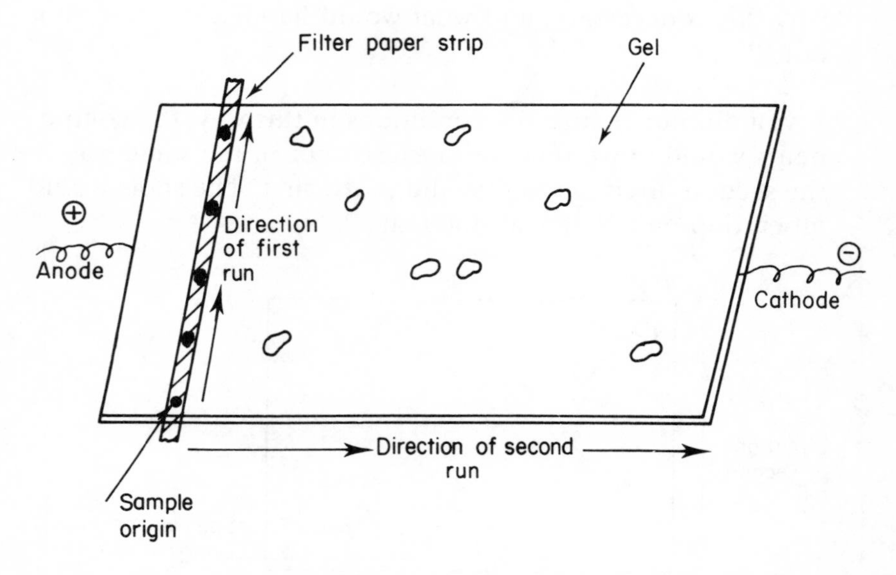

Fig. 7.2b. *Two-dimensional electrophoresis, different support media being used in each dimension*

The overall resolution of the mixture depends on the resolving power achieved in each dimension. Polyacrylamide gel electrophoresis, and particularly the SDS-polyacrylamide gel method, has a very high resolving power for protein separations, so many variations have been suggested for exploiting this in two-dimensional electrophoresis. The attraction of using SDS-polyacrylamide gels is that effects due to charge differences between protein molecules are eliminated, and their separation is on the basis of molecular size only (see Section 3.3.2b).

Now try this SAQ:

SAQ 7.2a

Using what you have learned in earlier sections of this Unit, suggest *three* possible ways to vary the conditions for two-dimensional electrophoresis.

7.3. COMBINED ELECTROPHORESIS AND ISOELECTRIC FOCUSSING

SDS-polyacrylamide gel electrophoresis and isoelectric focussing are the two most powerful techniques yet devised for the resolution of the components of mixtures of proteins. You may remember that isoelectric focussing separates protein molecules on the basis of their isoelectric points, whereas SDS-polyacrylamide gel electrophoresis separates them on the basis of their molecular size (see Sections 3.3.2b and 2.3.1). Consequently, these two techniques are ideally suited for use in combination for the two-dimensional separation of proteins.

The native proteins are first separated by isoelectric focussing (in polyacrylamide gel), and are then denatured *in situ* by incubating this gel in buffer containing SDS. The denatured proteins are transferred to an SDS-polyacrylamide slab gel, and electrophoresis is carried out. Usually, gels composed of a gradient of polyacrylamide concentration (and hence pore size – see Section 3.3.2) are used for the electrophoresis, to give a better spread of separated components.

It has been suggested that the resolving power of combined isoelectric focussing and SDS-polyacrylamide gel electrophoresis is sufficient to separate as many as 5000 components in a protein mixture (O'Farrell, 1975). This combination technique has been widely applied, particularly to the analysis of proteins in body fluids and cell extracts. Perhaps its most exciting application is in the study of gene products at the molecular level, to provide important information correlating the absence or presence of individual proteins characteristic of specific clinical conditions.

However, the combination technique of isoelectric focussing and SDS-polyacrylamide gel electrophoresis is not suitable for the quantitative analysis of proteins in a sample. There is often significant loss of proteins, particularly those of lower relative molecular mass, when the isoelectric focussing gel is treated with SDS, because of diffusion of the proteins out of the gel.

Summary

In this Part you have been introduced to the concept of two-dimensional techniques involving electrophoresis. Electrophoresis may be used in each of the two dimensions, or it may be used in only one dimension, in combination with either chromatography or isoelectric focussing in the other dimension. Examples have been presented of each of these possibilities.

Objectives

You should now be able to:

- discuss the advantages of two-dimensional techniques involving electrophoresis compared to one-dimension electrophoresis;

- outline the technique of protein fingerprinting (peptide mapping);

- interpret electrophoretograms obtained from protein fingerprinting;

- discuss homochromatography, two-dimensional electrophoresis, and combined electrophoresis–isoelectric focussing techniques.

We have now come to the end of this ACOL Unit on electrophoresis. I hope that you have found it an enjoyable and rewarding Unit. To gain full advantage of this text, it is advisable to try out the techniques of electrophoresis in supervised laboratory sessions.

References

Davis, B.J. (1964) *Ann. N.Y. Acad. Sci.*, **121**, 404

Ingram, V.M. (1957) *Nature*, **180**, 326

Karol, P.J. and Karol, M.H. (1978) *J. Chem. Educ.*, **55**, 626

Kohn, J. (1957) *Clin. Chim. Acta*, **2**, 297

Laas, T. (1972) *J. Chromatog.*, **66**, 347

Laurell, C.B. (1966) *Analyt. Biochem.*, **15**, 45

Maizel, J.V. (1966) *Science*, **151**, 988

O'Farrell, P.H. (1975) *J. Biol. Chem.*, **250**, 4007

Ornstein, L. (1964) *Ann. N.Y. Acad. Sci.*, **121**, 321

Poulik, M.D. and Smithies, O. (1958) *Biochem. J.*, **68**, 636

Righetti, P.G. and Drysdale, J.W. (1976) *Isoelectric Focusing*, North-Holland

Smithies, O. and Poulik, M.D. (1956) *Nature*, **177**, 1083

Tiselius, A. (1937) *Transactions of the Faraday Society*, **33**, 524

Watson, J.D. and Crick, F.H.C. (1953) *Nature*, **171**, 737

Weber, K. and Osborn, M. (1969) *J. Biol Chem.*, **244**, 4406

Self Assessment Questions and Responses

SAQ 1a

The general structure of the 20 amino acids (more correctly called 1-amino-1-carboxylic acids) commonly found in plant and animal proteins is

$$\begin{array}{l} R-CH-COOH \\ \quad\;\; | \\ \quad\;\, NH_2 \end{array}$$

where R represents one of 20 different groups. (Table 5 at the end of this Unit shows the full structure of these amino acids). Using this general formula, show how three amino acids are linked together to form a tripeptide.

Response

The three amino acids, $\underset{\displaystyle NH_2}{R'-\underset{|}{CH}-COOH}$, $\underset{\displaystyle NH_2}{R''-\underset{|}{CH}-COOH}$, and

$R'''-\underset{\displaystyle NH_2}{\underset{|}{CH}}-COOH$, are joined together by amide bonds, called *peptide bonds*, formed between the 1-amine group of one amino acid and the 1-carboxylic acid group of the next:

$$H_2N-\underset{\displaystyle R'}{\underset{|}{CH}}-\overset{\displaystyle O}{\overset{\|}{C}}-OH + H_2N-\underset{\displaystyle R''}{\underset{|}{CH}}-\overset{\displaystyle O}{\overset{\|}{C}}-OH + H_2N-\underset{\displaystyle R'''}{\underset{|}{CH}}-\overset{\displaystyle O}{\overset{\|}{C}}-OH$$

$$\Downarrow$$

$$H_2N-\underset{\displaystyle R'}{\underset{|}{CH}}-\boxed{\overset{\displaystyle O}{\overset{\|}{C}}-NH}-\underset{\displaystyle R''}{\underset{|}{CH}}-\boxed{\overset{\displaystyle O}{\overset{\|}{C}}-NH}-\underset{\displaystyle R'''}{\underset{|}{CH}}-C-OH + 3H_2O$$

I have indicated the peptide bonds, using boxes. Peptides, polypeptides and proteins are linear polymers of amino acids.

****

SAQ 1b

Although proline is listed in Table 5 as one of the 20 amino acids commonly found in plant and animal proteins, it is not a 1-amino-1-carboxylic acid. Write down the structural formula of the tripeptide in which proline is the central amino acid, linked to two others of structural formulae

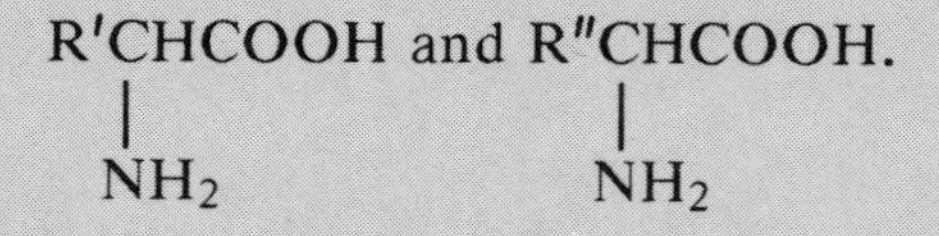

Response

Proline has a secondary amine group attached to the same carbon atom as its carboxyl group. The tripeptide you drew should have looked like this:

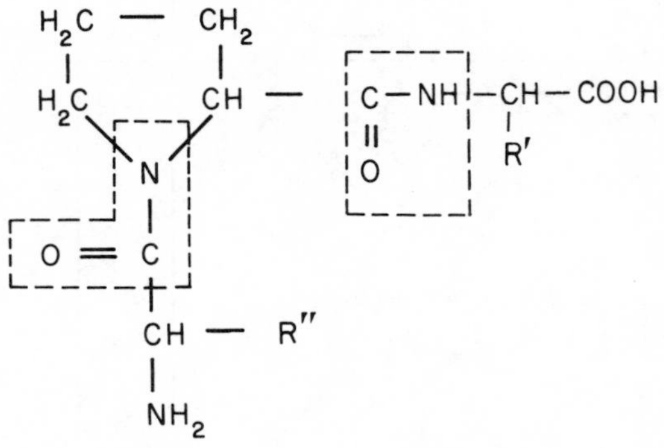

Again, I have outlined the two peptide bonds. Because of its unusual peptide bonding, there is usually a 'bend' in a polypeptide chain where proline is located.

****

SAQ 1c

> Name and draw the structural formulae of the major components produced on complete hydrolysis of (*i*) RNA and (*ii*) DNA.

Response

RNA and DNA molecules are polynucleotides which, on complete hydrolysis, give phosphoric acid, pentose sugar (either ribose or 2′-deoxyribose) and several heterocyclic bases, as shown in the following table:

	RNA	DNA
Phosphoric acid, H_3PO_4	✓	✓
Pentose sugar, D-ribose	✓	—
2'-deoxy-D-ribose	—	✓
Heterocyclic bases, purines: Adenine (A)	✓	✓
Guanine (G)	✓	✓
Pyrimidines: Cytosine (C)	✓	✓
Uracil (U)	✓	—
Thymine (T)	—	✓

****

SAQ 1d

> Draw a diagram to illustrate the polynucleotide structure of a single strand of RNA. Indicate how the structure of DNA differs from what you have drawn.

Response

There are several ways in which you could have drawn your diagram. The following is a 'general' diagram of a polynucleotide, where

B = heterocyclic base (A, G, C, U or T)

and P = phosphate.

The vertical lines represent sugar moieties, either ribose (in RNA) or 2′-deoxyribose (in DNA), and the sloping lines represent phosphodiester bonds between the 3′-hydroxyl group of one sugar and 5′-hydroxyl group of the next one.

In all RNA and DNA molecules, the phosphate groups form a 'backbone' from which the purine and pyrimidine bases protrude:

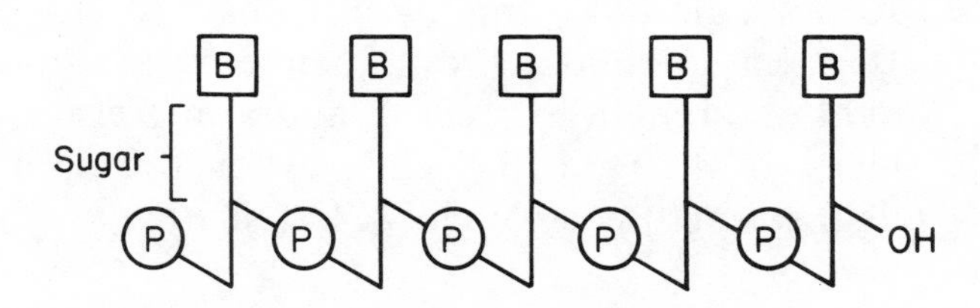

RNA molecules consist of a single polynucleotide strand, whereas DNA molecules (with exception of the DNA of some viruses) consist of two polynucleotide strands with hydrogen bonds between the purine and pyrimidine bases of the adjacent strands. The resultant structure for DNA molecules is the 'double helix' (first described by Watson and Crick, 1953):

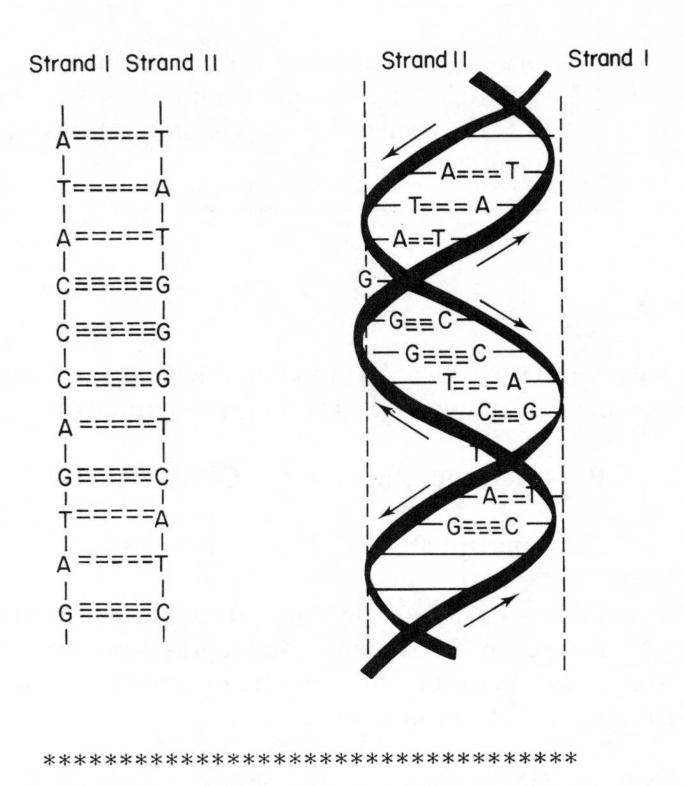

****

SAQ 1.1a

A molecule has a migration velocity of 1.60×10^{-3} cm s^{-1} under the influence of an applied field of strength 8 V cm^{-1}, and a migration velocity of 4.20×10^{-3} cm^{-1} when the applied field strength is 20 V cm^{-1}. Calculate:

(*i*) the electrophoretic mobility (μ) of the molecule, and

(*ii*) its migration velocity in a field of strength 12 V cm^{-1}.

Response

(*i*) Since $\mu = v/E$, where v is the migration velocity of a molecule under the influence of an applied electric field of strength E, we can calculate that:

$$\mu = v'/E' = \frac{1.60 \times 10^{-3}}{8} \text{ cm}^2 \text{ V s}^{-1}$$

$$= 0.20 \times 10^{-3} \text{ cm}^2 \text{ V s}^{-1}$$

$$\text{and } \mu'' = v''/E'' = \frac{4.20 \times 10^{-3}}{20} \text{ cm}^2 \text{ V s}^{-1}$$

$$= 0.21 \times 10^{-3} \text{ cm}^2 \text{ V s}^{-1}$$

from which,

$$\mu = 0.205 \times 10^{-3} \text{ cm}^2 \text{ V s}^{-1}$$

(*ii*) Substituting this value for μ in the equation,

$$\mu = v'''/E''', \text{ where } E''' = 12 \text{ V cm}^{-1}$$

$$\text{we find } v''' = 12 \times 0.205 \times 10^{-3} \text{ cm s}^{-1}$$

$$= 2.46 \times 10^{-3} \text{ cm s}^{-1}$$

ie the migration velocity of the molecule under an applied field of strength 12 V cm^{-1} should be 2.46×10^{-3} cm s^{-1}.

The calculations here are quite straightforward, so hopefully they did not give you much trouble. However, it is also important that you used the correct *units* for migration velocity, field strength and electrophoretic mobility.

****

SAQ 1.2a

> By referring to Table 5, write out a list of the amino acids that have ionisable groups in their side chains, giving the pK_a value for each of them.

Response

For this SAQ, your response should have been:

Amino Acid	pK_a of side-chain
aspartic acid	3.86 (2-carboxyl)
glutamic acid	4.25 (3-carboxyl)
lysine	10.53 (5-amino)
arginine	12.48 (guanidino)
histidine	6.00 (imidazole)
tyrosine	10.07 (phenol)
cysteine	8.33 (sulphydryl)

Aspartic acid and glutamic acid each have a carboxyl group in their side-chain, whereas lysine and arginine each have a side-chain amine group.

The other amino acids are more varied. The imidazole group present in histidine readily loses or gains a proton in aqueous solution, depending on the nature of other ionisable groups in its environment:

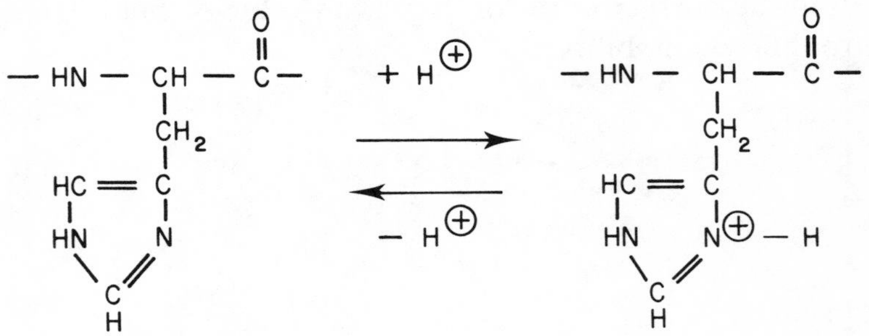

```
                 O                                         O
                 ||                                        ||
— HN — CH —  C —        + H⊕             — HN — CH —  C —
        |                                         |
        CH2             ———————→                  CH2
        |                                         |
  HC =  C               ←———————           HC =  C
  |     |                                   |     |
  HN    N               − H⊕                HN    N⊕ — H
    \  //                                     \  //
     C                                         C
     H                                         H
```

The phenolic hydrogen of tyrosine can dissociate to give a phenoxide-type anion:

$$-HN-CH(CH_2C_6H_4OH)-C(=O)- \underset{+H^{\oplus}}{\overset{-H^{\oplus}}{\rightleftharpoons}} -HN-CH(CH_2C_6H_4O^{\ominus})-C(=O)-$$

and the sulphydryl group of cysteine can lose its hydrogen under fairly mild conditions:

$$-HN-CH(CH_2SH)-C(=O)- \underset{+H^{\oplus}}{\overset{-H^{\oplus}}{\rightleftharpoons}} -HN-CH_2(CH_2S^{\ominus})-C(=O)-$$

****

SAQ 1.2b

The table below lists a number of proteins and their isoelectric points (pI). For each protein, decide whether its net charge will be positive or negative at (*i*) pH 3.0. (*ii*) pH 7.4. and (*iii*) pH 10.0 ⟶

SAQ 1.2b (cont.)

Protein	pI	Net charge		
		pH 3.0	pH 7.4	pH 10.0
collagen	6.7			
serum albumin	4.8			
lysozyme	11.1			
human haemoglobin	7.1			
insulin	5.4			
cytochrome c	10.0			
pepsin	1.0			

Response:

Your completed table should have looked like this.

Protein	pI	Net charge		
		pH 3.0	pH 7.4	pH 10.0
collagen	6.7	+ve	−ve	−ve
serum albumin	4.8	+ve	−ve	−ve
lysozyme	11.1	+ve	+ve	+ve
human haemoglobin	7.1	+ve	−ve	−ve
insulin	5.4	+ve	−ve	−ve
cytochrome c	10.0	+ve	+ve	zero
pepsin	1.0	−ve	−ve	−ve

The rule to remember for this SAQ is that at pH values below its isoelectric point a protein will have more of its side-chain groups protonated than not, and it will therefore have a net positive charge. Conversely, at pH values higher than its isoelectric point the majority of the side-chain groups of a protein will be dissociated, and it will therefore have a net negative charge. At its isoelectric point a protein will have no net charge, as an equal number of its side-chain groups will be protonated and dissociated, and this is why cytochrome c has no net charge at pH 10.0, its isoelectric point.

Pepsin, which has an exceptionally low pI for a protein, is the principal proteolytic enzyme in the gastric juice in all vertebrates. Since the pH of gastric juice is near 1 or 2, pepsin must possess unusual stability in acid solutions.

****

SAQ 1.2c

From the response to SAQ 1.2b, what do you suggest would be a suitable pH of solution to use for the optimum separation of human haemoglobin and cytochrome c – pH 3.0, pH 7.4, or pH 10.0? Give a reason for your answer.

Response:

The answer to this is pH 7.4, since at this pH human haemoglobin carries a net negative charge, whereas cytochrome c carries a net positive charge, so they should migrate towards opposite electrodes under electrophoresis. However, pH 7.4 is very close to the isoelectric point of human haemoglobin, so its net charge will be very small at this pH and its migration towards the anode may be affected significantly by electro-osmotic effects. Perhaps a better pH to use would be one greater than 7.4, but less than 10.0, say around 8.5.

****

SAQ 2.2a

> Make a list of the properties that you think would be desirable in a substance which is to be used as support material in zone electrophoresis.

Response

There are a number of properties that you could have chosen. The following are probably the most important, so if you have mentioned many or all of them you are certainly on the right track.

Support material should:

(*i*) have *no ionisable groups*, otherwise it would take on a net charge when an external electric field was applied to it. This could have two effects. The charged groups on the support medium could interact with the charged molecules in the sample and interfere with their migration. Charged groups on the support medium could also lead to an electro-osmotic effect.

(*ii*) be *chemically inert*, so that it will not interact with either the samples or the buffer under the electrophoretic conditions.

(*iii*) be *strong when wet*, otherwise it might fall apart during the electrophoresis.

(*iv*) be *homogeneous* in both chemical and physical properties, so that the migration of substances is not affected by variations or discontinuities in the support medium.

(*v*) be *inexpensive* – cost must be a factor in selection of a suitable support medium.

(*vi*) not get too warm (*overheat*) during electrophoresis. Many electrophoretic systems incorporate a means of cooling. Overheating leads to distortion of the zones, and may even lead to denaturation of the samples. (see Section 4.2.5).

(*vii*) not interfere with the means of detection used for locating separated components. It is important that the 'background' of the detection methods is as low as possible.

****

SAQ 3.1a

To summarise what we have been discussing in this section, try writing out two lists, one of the advantages and the other of the disadvantages of paper as a medium for zone electrophoresis.

Response

There are a number of possible advantages and disadvantages that you might have mentioned. The following is a comprehensive list – if you have mentioned most of these points then you have done well:

Advantages:

(*i*) equipment required is relatively simple and inexpensive,

(*ii*) the technique is simple to perform,

(*iii*) the electrophoretic run is usually quite quick – an hour or two,

(*iv*) the voltage required is low – around 20 $V\ cm^{-1}$,

(*v*) it is a technique that is applicable to a wide range of molecules,

(*vi*) the detection of separated components is easy,

(*vii*) the stained paper can be kept as a permanent record of the results.

Disadvantages:

(*i*) there is liable to be variability in the quality of the paper, from batch to batch, and also perhaps even within one sheet,

(*ii*) some papers have adsorptive properties, which could interfere with the electrophoretic mobility of sample molecules,

(*iii*) ionisable groups on the paper can lead to an electro-osmotic effect which would distort the migration characteristics of the separating components,

(*iv*) the effects of evaporation of water from the paper during the electrophoretic run can change the buffer composition,

(*v*) heating effects can lead to diffusion of the zones, so a system of cooling is desirable – but it is not always practicable,

(*vi*) generally, there is quite significant diffusion of the separated zones, which means that the resolution of paper electrophoresis is not particularly good, especially for complex mixtures,

(*vii*) the low voltage means that low relative molecular mass components are poorly separated,

(*viii*) it is not readily adapted to preparative-scale separations.

****

SAQ 3.3a

The proteins myoglobin and ovalbumin have relative molecular masses (M_r), 17,200 and 43,000 respectively. During electrophoresis through a polyacrylamide gel, they migrate 1.50 cm and 5.50 cm respectively. What is the M_r of the protein, α-chymotrypsinogen, that migrates 3.25 cm in the same gel?

Response

To answer this SAQ you should use Eq. 3.1:

$$D = a - b.\log_{10}M$$

For myoglobin,

$D = 1.50$ and $M = 17{,}200$, so $1.50 = a - b.\log_{10} 17{,}200$

$= a - 4.24b$

For ovalbumin,

$D = 5.50$ and $M = 43{,}000$, so $5.50 = a - b.\log_{10} 43{,}000$

$= a - 4.63b$

We now have a pair of simultaneous equations.

From the first of these equations, $a = 1.50 + 4.24b$, and if we substitute this value for a in the second equation we get:

$$5.50 = 1.50 + 4.24b - 4.63b$$

from which $4.00 = -0.39b$, so $b = -4.00/0.39 = -10.3$.

If we now use this value for b, we can deduce that

$$a = 1.50 + 4.24 \times -10.3$$

so that $a = 1.50 - 43.67$

$= -42.17.$

If we substitute both these values, for a and b, in Eq. 3.1 we can determine the M_r of α-chymotrypsinogen, in the following way, D being 3.25:

$$3.25 = a - b.\log_{10}M$$
$$3.25 = -42.17 - (-10.3)\log_{10}M$$

from which $\log_{10}M = (3.25 + 42.17)/10.3$
$= 45.42/10.3$
$= 4.41$
and so $M = 25{,}700$

That is, the relative molecular mass of α-chymotrypsinogen is 25,700.

****

SAQ 3.3b

> Give *three* reasons why the rate of migration of a polypeptide–SDS complex in SDS–polyacrylamide gel electrophoresis depends on its size.

Response

You should have mentioned the following points in your answer:

(*i*) The net charge of a polypeptide–SDS complex is determined entirely by the amount of SDS bound to it, and not at all by the protonation of its amino acids. The larger the polypeptide (or protein), the more SDS that will bind to it (1.4 g SDS per g protein), and hence the higher its net charge.

(*ii*) The migration of a polypeptide–SDS complex through a sieving gel, such as polyacrylamide, depends on its size because molecular complexes larger than the average pore size of the gel will be held back more than smaller ones which can pass more easily through the gel pores (see Fig. 3.3a).

(*iii*) Separation of molecules through sieving gels depends on their shape as well as their size. However, treatment of polypeptides with SDS produces complexes which have a random-coil shape, so their shapes are all similar.

****

SAQ 3.3c

Indicate whether each of the following statements is TRUE or FALSE:

	TRUE	FALSE
(*i*) In polyacrylamide gels, the higher the concentration of polymer the smaller the average pore size of the gel		
(*ii*) A single band in polyacrylamide disc gel electrophoresis proves that the sample consists of only one component		
(*iii*) For the accurate determination of the relative molecular mass of a polypeptide, the method of choice is disc gel electrophoresis		

⟶

SAQ 3.3c (cont.)

	TRUE	FALSE
(*iv*) Agarose gel electrophoresis is useful for the separation of molecules that are too large to enter the pores of polyacrylamide gels		
(*v*) The relative molecular masses of RNA molecules can be determined using polyacrylamide gel electrophoresis		

Response:

(*i*) This is *true*, and can be used to tailor gel pore size to the sizes of the molecules to be separated.

(*ii*) This is *false*. A single band may indicate that only one component is present, but it certainly does not prove it. In disc electrophoresis, the molecules are separated according to both their size and their net charge, therefore more than one component may be present in a single 'band', especially if the band is rather broad.

(*iii*) No, this is *false*. The method of choice would be SDS-polyacrylamide gel electrophoresis.

(*iv*) *True* – agarose gels have been very useful for the electrophoretic separation of large macromolecules, especially double-stranded DNA.

(*v*) Yes, this is *true*. Polyacrylamide gels, particularly slab gels, have been very useful for this type of study.

If you managed to get all these answers correct, you clearly have a good grasp of the main points of Part 3.

****

SAQ 4.2a

For each of the purposes (*i*)–(*v*), decide on an appropriate electrophoretic support medium, and enter your choice on the lines provided:

	Purpose	Support Medium
(*i*)	analysis of a mixture of DNA molecules of M_r 50 to 100 × 10^6	________
(*ii*)	analysis of a sample of glutamic acid to check for its contamination with glutamine	________
(*iii*)	accurate determination of the relative molecular masses of the two polypeptide chains of human insulin	________

⟶

SAQ 4.2a (cont.)

Purpose	Support Medium
(*iv*) separation of the isoenzymes of lactic dehydrogenase (eg from the serum of a patient with a myocardial infraction)	______
(*v*) determination of the number of different coat proteins of a virus	______

Response

Purpose	Support Medium
(*i*) analysis of a mixture of DNA molecules of M_r 50–100 × 10^6	agarose gel
(*ii*) analysis of a sample of glutamic acid to check for its contamination with glutamine	filter paper or cellulose acetate
(*iii*) accurate determination of the relative molecular masses of the two polypeptide chains of human insulin	SDS – polyacrylamide gel
(*iv*) separation of the isoenzymes of lactic dehydrogenase (eg from the serum of a patient with a myocardial infraction)	starch gel or cellulose acetate
(*v*) determination of the number of different coat proteins of a virus	polyacrylamide disc gel

My reasons for these choices are as follows:

(*i*) Double-stranded DNA molecules of this size range are too large to pass through the pores of polyacrylamide gels. Agarose gels can be made that have pore sizes large enough to allow these large molecules to enter the gel and yet still be rigid enough to be useful as a support medium.

(*ii*) If you refer to Table 5 you will see that glutamic acid has a carboxylic acid group on its side-chain, and its isoelectric point (pI) is 3.22. Its relative molecular mass (147) is almost the same as that of glutamine (146), whose side-chain has an amide group in it, and whose isoelectric point is 5.65. Between pH 3.22 and pH 5.65, glutamic acid will be negatively charged whereas glutamine will be positively charged, so they could be separated by electrophoresis in a buffer of pH 3.22–5.65. Above pH 5.65, both these amino acids will be negatively charged, but glutamic acid will always be more negatively charged than glutamine, and will migrate faster towards the anode during electrophoresis. These two amino acids can be readily separated using a simple electrophoresis system, such as paper or cellulose acetate.

(*iii*) SDS-polyacrylamide gel electrophoresis is used to separate polypeptide chains of samples that have been treated with a mixture of SDS and 2-mercaptoethanol, which breaks any disulphide bridges. The two polypeptide chains of human insulin are held together by 3 disulphide bridges. Migration of SDS-polypeptide complexes depends only on their molecular size, so this technique is useful for the accurate determination of relative molecular mass.

(*iv*) Starch gel electrophoresis is still used for the rapid analysis of isoenzymes. The resolution of the technique is good enough to distinguish the presence or absence of individual isoenzymes in a sample. Many hospital laboratories are using cellulose acetate for this purpose, which is easier and quicker to use than starch gels.

(*v*) Disc polyacrylamide gel electrophoresis is an excellent technique for deciding how many different protein components are in a sample. The individual proteins show up as clearly defined disc-shaped bands after electrophoresis. However, to be certain about the number of proteins present, it is advisable to repeat the analysis using a variety of buffer pH values, and/or a variety of different gel concentrations (ie pore sizes).

****

SAQ 4.2b

In determining the conditions for the maximum electrophoretic separation of two proteins, which parameter is likely to have the *greatest* effect?

Choose one of the following:

(*i*) pH
(*ii*) ionic strength
(*iii*) temperature
(*iv*) current

Response

The correct response to this SAQ is (*i*) pH.

Although all of these parameters affect the electrophoretic mobility and hence the separation of proteins, some will have similar effects on all the proteins in the sample.

pH has a very significant effect on the magnitude of the charge carried by a protein molecule. Since pH affects the protonation and ionisation of the side-chains of the amino acids of which a protein is composed, it affects different proteins to different extents, because the amino acid composition and sequence of every protein is unique.

Perhaps you thought *ionic strength* would have the greatest effect on the separation? If so, you are right in realising that the ionic strength also affects the charge characteristics of protein molecules, since charged macromolecules become surrounded by a 'shield' of counterions which effectively decreases the net charge of the molecule and so affects its electrophoretic mobility. The size of this effect depends on the nature and charges of the ionisable groups on the surface of the macromolecule. So the effect is likely to be different for each protein in a sample. However, the effects of ionic strength are not as great as those of pH, which affects amino acid side-chains inside the molecules as well as those on their surface.

Changes in *temperature* can have very significant effects on the electrophoretic separation of protein molecules. As we have discussed in Section 4.2.5, temperature can alter several of the parameters of an electrophoretic system. But these changes are not highly specific for individual protein molecules, and all the molecules in a sample will be affected in much the same way, although there will be small individual differences.

Changing the *current* will generally have similar effects on all the protein molecules in a sample, although individual molecules may be affected slightly differently.

****

SAQ 4.2c

In the electrophoretic separation of several molecules of double-stranded DNA, which parameter is likely to have the *greatest* effect?

Choose one of the following:

(*i*) pH
(*ii*) buffer ionic composition
(*iii*) temperature
(*iv*) gel pore size

Response

In this case I hope you did not choose pH! The correct response is (*iv*), gel pore size.

You may remember from Section 1.2.2 that all double stranded DNA molecules are negatively charged in solution, because of the acidic nature of their phosphate backbone. Changing the *pH* may change the size of this negative charge slightly, but it will affect all the DNA molecules in the sample in the same way, so it will have little or no effect on the separation of these molecules.

The *composition of the buffer* could have an effect on the charge of the DNA molecules, since different ions in the buffer could interact differently with the negatively charged DNA molecules. But again the effect would be the same with all the DNA molecules.

Changing the *temperature* can alter several of the parameters of the electrophoretic system, some of which could change the electrophoretic mobility of the DNA molecules – but again they would all be affected in the same way.

However, because DNA molecules all have the same charge-to-size ratios, the sieving effect of different gels is exploited to maximise the separation of DNA molecules of different relative molecular masses. The resolution of these techniques is superb – by carefully choosing an appropriate *pore size*, DNA molecules of even quite similar sizes can be clearly separated from one another.

****

SAQ 5.1a

Select from the stains listed below the one that is most appropriate for each of the purposes (*i*)–(*vi*), and enter your choice in the spaces provided:

(*i*) staining of proteins in polyacrylamide gels __________

(*ii*) detection of small peptides after paper electrophoresis __________

(*iii*) fluorescent labelling of double stranded DNA molecules __________

(*iv*) identification of bands of glycoproteins after electrophoresis of a mixture of proteins __________

(*v*) rapid detection of proteins separated on cellulose acetate membranes __________

(*vi*) fluorescent tagging of proteins in SDS-polyacrylamide gels __________

Alcian Blue
Amido Black
Coomassie Brilliant Blue
Ethidium Bromide
Fluorescamine
Ninhydrin
Ponceau S
Pyronine Y

Response

(*i*)	staining of proteins in polyacrylamide gels	Coomassie Blue
(*ii*)	detection of small peptides after paper electrophoresis	Ninhydrin
(*iii*)	fluorescent labelling of double stranded DNA molecules	Ethidium Bromide
(*iv*)	identification of bands of glycoproteins after electrophoresis of a mixture of proteins	Alcian Blue
(*v*)	rapid detection of proteins separated on cellulose acetate membranes	Ponceau S
(*vi*)	fluorescent tagging of proteins in SDS-polyacrylamide gels	Fluorescamine

If you made the choices shown above, well done! The information can all be found in Section 5.1.2.

(*i*) Proteins in polyacrylamide gels may be detected colorimetrically by staining them with either Amido Black or Coomassie Brilliant Blue, but the use of 7% ethanoic acid with Amido Black leads to some dehydration and shrinkage of polyacrylamide gels, which can make accurate measurement of mobility difficult. So the preferred stain is Coomassie Brilliant Blue.

(*ii*) Small peptides are readily detected after electrophoresis on paper with the reagent ninhydrin, which produces the dye Ruhemann's purple:

Ninhydrin + $NH_2-CH(R)-COOH \rightarrow$ Hydrindantin + $RCHO + CO_2 + NH_3$

Hydrindantin + Ninhydrin $\rightarrow$ Ruhemann's purple

(*iii*) Double-stranded polynucleotides, such as DNA or small pieces of double-stranded RNA, react with the fluorescent dye ethidium bromide to form highly fluorescent complexes, which are readily detectable under a UV lamp.

(*iv*) Compound proteins, such as glycoproteins, can be distinguished from simple proteins by stains specific for the non-protein part of the molecule. In the case of glycoproteins the sugar moiety can be stained by Alcian Blue.

(*v*) After electrophoresis on cellulose acetate strips, proteins are quickly and clearly detected by the pink-staining reagent Ponceau S.

(*vi*) And finally, to detect proteins in SDS-polyacrylamide gels it is often useful to tag them before electrophoresis using one of a number of fluorescent dyes that are specific for proteins – for example fluorescamine.

(Pyronine Y is a fluorescent dye used to tag polynucleotides).

****

SAQ 5.2a

From the radioisotopes given below, choose the one you think would be the *most* appropriate for rapid assay of the various samples after gel electrophoresis. Put your choice on the lines provided.

(*i*) Proteins containing several methionine residues __________

(*ii*) Newly-synthesised RNA molecules __________

(*iii*) Proteins containing the amino acid tyrosine __________

(*iv*) New proteins synthesised in cells following viral infection __________

^{35}S ^{32}P ^{125}I ^{3}H ^{14}C

Response

(*i*) Proteins containing several methionine residues ^{35}S

(*ii*) Newly-synthesised RNA molecules ^{32}P

(*iii*) Proteins containing the amino acid tyrosine ^{125}I

(*iv*) New proteins synthesised in cells following viral infection ^{14}C

The correct responses are shown above.

(*i*) As you can see from Table 5, the amino acid methionine contains one atom of sulphur. This amino acid is commercially available labelled with ^{35}S, and if proteins are synthesised in the presence of ^{35}S-labelled methionine the ^{35}S label will be incorporated into the proteins. This provides a radiochemical tag for those proteins that contain methionine. Certainly methionine labelled with other radioisotopes could be used, such as ^{14}C or ^{3}H, but ^{35}S would probably be the isotope of choice.

(*ii*) Ribonucleic acids are synthesised from nucleotide precursors, which contain phosphate groups, one phosphate group being incorporated into the nucleic acid for each base incorporated. Consequently, a very useful way of tagging newly synthesised nucleic acids is to use nucleotide precursors labelled with ^{32}P, which is a high energy isotope that is easily detected.

(*iii*) ^{125}I can be used to specifically label tyrosine or lysine residues in proteins, depending on the actual technique used. ^{125}I is a readily detected γ-emitter.

For question (*iv*), the answer is less clear cut. You actually have a choice of isotopes to use – a large number of radioactively labelled amino acids are available commercially. Those labelled with ^{35}S will only tag proteins containing either methionine or cysteine residues. However, if you choose a radioactively labelled, commonly-occurring amino acid such as leucine or phenylalanine, that is not readily broken down by cells to other non-amino acid metabolites, a much greater number of proteins will be labelled. In this way you would be more likely to label all of the proteins synthesised after viral infection. Remember that ^{14}C is more easily detected than ^{3}H, because it emits higher energy radiation.

****

SAQ 5.2b

Suggest *three* factors which would affect the sensitivity of detection of radioactivity in labelled molecules after electrophoresis.

Response

The points you may have thought of should have included the following:

(*i*) the method of detection used eg autoradiography, scintillation counting

(*ii*) the nature and energy of the emitted radiation.

(*iii*) the properties of the support medium eg its composition, thickness etc.

****

SAQ 6.2a

In Fig. 6.2a, the standard reservoirs, (*a*)–(*e*), contained 1.0, 0.75, 0.5, 0.25 and 0.1 μg of antigen respectively. Draw a calibration graph relating rocket height to antigen amount, and from it determine the concentration of this antigen in the sample.

Response

First, use your ruler to measure the distance in mm from the upper edge of the standard wells to the tips of their corresponding rockets.

Your standard graph is drawn from the following information:

Mass of standard antigen (μg)	Height of rocket (mm)
1.00	42
0.75	31
0.50	21.5
0.25	13
0.10	6.5

and it should look like this:

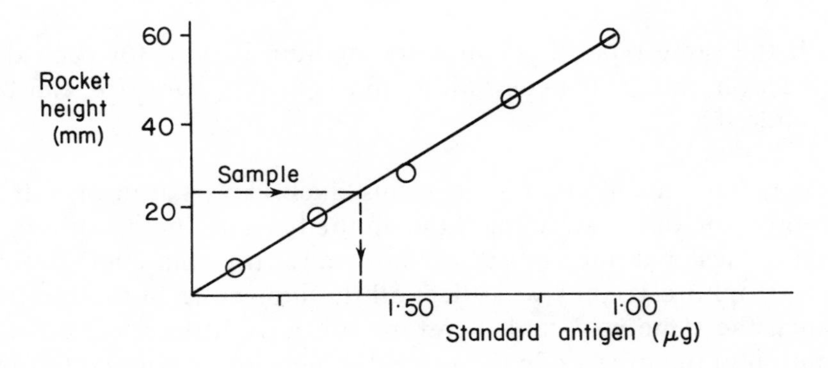

Now measure the heights of the sample rockets. These are 18 mm, 17.5 mm and 19 mm, with an average height of 18.2 mm.

On the standard graph, a rocket height of 18.2 mm corresponds to an amount of antigen of 0.4 μg, therefore 0.4 μg antigen is present in 5 μl of sample. The concentration of antigen in the sample is 0.4 μg per 5 μl, or 0.08 mg cm^{-3}.

****

SAQ 7.2a

> Using what you have learned in earlier sections of this Unit, suggest *three* possible ways to vary the conditions for two-dimensional electrophoresis.

Response

You should have suggested the following possibilities:

(*i*) Changing the *pH* value of the buffer. The electrophoretic mobilities of the components in the mixture depend on their net charge, which in turn depends on the buffer pH.

(*ii*) Using a chemically different *support medium* for each direction

(*iii*) If the same type of gel support medium is used for each direction, the gel concentration, and hence its *pore size*, can be altered.

You may have made other suggestions, such as variation of buffer concentration, or the strength of the applied electric field. Although changing these parameters will certainly alter the characteristics of the electrophoresis, they will affect all the molecules in the sample in much the same way, and therefore will have little effect on the resolution of the mixture in the second dimension compared to that achieved in the first dimension.

Units of Measurement

For historic reasons a number of different units of measurement have evolved to express quantity of the same thing. In the 1960s, many international scientific bodies recommended the standardisation of names and symbols and the adoption universally of a coherent set of units—the SI units (Système Internationale d'Unités)—based on the definition of five basic units: metre (m); kilogram (kg); second (s); ampere (A); mole (mol); and candela (cd).

The earlier literature references and some of the older text books, naturally use the older units. Even now many practicing scientists have not adopted the SI unit as their working unit. It is therefore necessary to know of the older units and be able to interconvert with SI units.

In this series of texts SI units are used as standard practice. However in areas of activity where their use has not become general practice, eg biologically based laboratories, the earlier defined units are used. This is explained in the study guide to each unit.

Table 1 shows some symbols and abbreviations commonly used in analytical chemistry; Table 5 is a list of amino acids commonly found in proteins. Table 2 shows some of the alternative methods for expressing the values of physical quantities and the relationship to the value in SI units.

More details and definition of other units may be found in the *Manual of Symbols and Terminology for Physicochemical Quantities and Units*, Whiffen, 1979, Pergamon Press.

Table 1 *Symbols and Abbreviations Commonly used in Analytical Chemistry*

Symbol	Meaning
Å	Angstrom
$A_r(X)$	relative atomic mass of X
A	ampere
E or U	energy
G	Gibbs free energy (function)
H	enthalpy
J	joule
K	kelvin (273.15 + t °C)
K	equilibrium constant (with subscripts p, c, therm etc.)
K_a, K_b	acid and base ionisation constants
$M_r(X)$	relative molecular mass of X
N	newton (SI unit of force)
P	total pressure
s	standard deviation
T	temperature/K
V	volume
V	volt (J A^{-1} s^{-1})
a, a(A)	activity, activity of A
c	concentration/ mol dm^{-3}
e	electron
g	gramme
i	current
s	second
t	temperature / °C
bp	boiling point
fp	freezing point
mp	melting point
$\approx$	approximately equal to
$<$	less than
$>$	greater than
e, exp(x)	exponential of x
ln x	natural logarithm of x; $\ln x = 2.303 \log x$
log x	common logarithm of x to base 10

Table 2 *Alternative Methods of Expressing Various Physical Quantities*

1. **Mass (SI unit : kg)**

$$g = 10^{-3}\ kg$$
$$mg = 10^{-3}\ g = 10^{-6}\ kg$$
$$\mu g = 10^{-6}\ g = 10^{-9}\ kg$$

2. **Length (SI unit : m)**

$$cm = 10^{-2}\ m$$
$$Å = 10^{-10}\ m$$
$$nm = 10^{-9}\ m = 10Å$$
$$pm = 10^{-12}\ m = 10^{-2}\ Å$$

3. **Volume (SI unit : m^3)**

$$l = dm^3 = 10^{-3}\ m^3$$
$$ml = cm^3 = 10^{-6}\ m^3$$
$$\mu l = 10^{-3}\ cm^3$$

4. **Concentration (SI units : mol m^{-3})**

$$M = mol\ l^{-1} = mol\ dm^{-3} = 10^3\ mol\ m^{-3}$$
$$mg\ l^{-1} = \mu g\ cm^{-3} = ppm = 10^{-3}\ g\ dm^{-3}$$
$$\mu g\ g^{-1} = ppm = 10^{-6}\ g\ g^{-1}$$
$$ng\ cm^{-3} = 10^{-6}\ g\ dm^{-3}$$
$$ng\ dm^{-3} = pg\ cm^{-3}$$
$$pg\ g^{-1} = ppb = 10^{-12}\ g\ g^{-1}$$
$$mg\% = 10^{-2}\ g\ dm^{-3}$$
$$\mu g\% = 10^{-5}\ g\ dm^{-3}$$

5. **Pressure (SI unit : N m^{-2} = kg m^{-1} s^{-2})**

$$Pa = Nm^{-2}$$
$$atmos = 101\ 325\ N\ m^{-2}$$
$$bar = 10^5\ N\ m^{-2}$$
$$torr = mmHg = 133.322\ N\ m^{-2}$$

6. **Energy (SI unit : J = kg m^2 s^{-2})**

$$cal = 4.184\ J$$
$$erg = 10^{-7}\ J$$
$$eV = 1.602 \times 10^{-19}\ J$$

Table 3 *Prefixes for SI Units*

Fraction	Prefix	Symbol
10^{-1}	deci	d
10^{-2}	centi	c
10^{-3}	milli	m
10^{-6}	micro	μ
10^{-9}	nano	n
10^{-12}	pico	p
10^{-15}	femto	f
10^{-18}	atto	a

Multiple	Prefix	Symbol
10	deka	da
10^{2}	hecto	h
10^{3}	kilo	k
10^{6}	mega	M
10^{9}	giga	G
10^{12}	tera	T
10^{15}	peta	P
10^{18}	exa	E

Table 4 *Recommended Values of Physical Constants*

Physical constant	Symbol	Value
acceleration due to gravity	g	9.81 m s^{-2}
Avogadro constant	N_A	$6.022\,05 \times 10^{23} \text{ mol}^{-1}$
Boltzmann constant	k	$1.380\,66 \times 10^{-23} \text{ J K}^{-1}$
charge to mass ratio	e/m	$1.758\,796 \times 10^{11} \text{ C kg}^{-1}$
electronic charge	e	$1.602\,19 \times 10^{-19} \text{ C}$
Faraday constant	F	$9.648\,46 \times 10^{4} \text{ C mol}^{-1}$
gas constant	R	$8.314 \text{ J K}^{-1} \text{ mol}^{-1}$
'ice-point' temperature	T_{ice}	273.150 K exactly
molar volume of ideal gas (stp)	V_m	$2.241\,38 \times 10^{-2} \text{ m}^3 \text{ mol}^{-1}$
permittivity of a vacuum	ϵ_o	$8.854\,188 \times 10^{-12} \text{ kg}^{-1} \text{ m}^{-3} \text{ s}^4 \text{ A}^2 \text{ (F m}^{-1})$
Planck constant	h	$6.626\,2 \times 10^{-34} \text{ J s}$
standard atmosphere pressure	p	$101\,325 \text{ N m}^{-2}$ exactly
atomic mass unit	m_u	$1.660\,566 \times 10^{-27} \text{ kg}$
speed of light in a vacuum	c	$2.997\,925 \times 10^{8} \text{ m s}^{-1}$

Table 5 *The twenty amino acids commonly found in proteins*

Name	Abbreviation	Formula	Values of PK_a
Glycine	Gly	$N_2N.CH_2.COOH$	2.34; 9.60
Alanine	Ala	$H_2N.CH.COOH$ | CH_3	2.35; 9.69
Valine	Val	$H_2N.CH.COOH$ | CH | $(CH_3)_2$	2.32; 9.62
Leucine	Leu	$H_2N.CH.COOH$ | CH_2 | CH | $(CH_3)_2$	2.36; 9.60
Isoleucine	Ile	$H_2N.CH.COOH$ | $CH.CH_3$ | $CH_2.CH_3$	2.36; 9.68
Proline	Pro	$CH_2-CH.COOH$ | CH_2 NH CH_2	1.99; 10.60
Phenylaline	Phe	$H_2N.CH.COOH$ | CH_2 |	1.83; 9.13
Tryptophane	Trp	$H_2N.CH.COOH$ | CH_2 | | N–H	2.38; 9.39

Table 5 (cont.)

Name	Abbreviation	Formula	Values of PK_a
Methionine	Met	$H_2N.CH.COOH$ $\vert$ CH_2 $\vert$ CH_2 $\vert$ S $\vert$ CH_3	2.28; 9.21
Aspartic Acid	Asp	$H_2N.CH.COOH$ $\vert$ CH_2 $\vert$ $COOH$	2.09 (α-carboxyl) 3.86 (β-carboxyl) 9.82
Glutamic Acid	Glu	$H_2N.CH.COOH$ $\vert$ CH_2 $\vert$ CH_2 $\vert$ $COOH$	2.19 (α-carboxyl) 4.25 (γ-carboxyl) 9.67
Lysine	Lys	$H_2N.CH.COOH$ $\vert$ $(CH_2)_4$ $\vert$ NH_2	2.18 8.95 (α-amino) 10.53 (ϵ-amino)
Arginine	Arg	$H_2N.CH.COOH$ $\vert$ $(CH_2)_3$ $\vert$ NH $\vert$ $C{=}NH$ $\vert$ NH_2	2.17 9.04 (α-amino) 12.48 (guanidine)
Histidine	His	$H_2N.CH.COOH$ $\vert$ CH_2 $\vert$ $C{=}CH$ $\vert$ $HN \quad N$ CH	1.82 6.0 (imidazole) 9.17
Serine	Ser	$H_2N.CH.COOH$ $\vert$ CH_2OH	2.21; 9.15

Table 5 (cont.)

Name	Abbreviation	Formula	Values of PK_a
Threonine	Thr	$H_2N.CH(CHOH.CH_3).COOH$	2.63; 10.43
Tyrosine	Tyr	$H_2N.CH(CH_2-C_6H_4-OH).COOH$	2.20 9.11 (α-amino) 10.07 (phenolic hydroxyl)
Cysteine	Cys	$H_2N.CH(CH_2SH).COOH$	1.71 8.33 (sulphydryl) 10.78 (α-amino)
Asparagine	Asn	$H_2N.CH(CH_2.C{=}O.NH_2).COOH$	2.02; 8.8
Glutamine	Gln	$H_2N.CH((CH_2)_2.C{=}O.NH_2).COOH$	2.17; 9.13